GREEN PROCESS OF THREE-COMPONENT PROSTAGLANDIN SYNTHESIS AND RAPID 11 C LABELINGS FOR SHORT-LIVED PET TRACERS

Authored by **Masaaki Suzuki, Hiroko Koyama, Hideki Ishii, Koichi Kato, Margareta Ögren** and **Hisashi Doi**

Green Process of Three-Component Prostaglandin Synthesis and Rapid 11 C Labelings for Short-Lived PET Tracers
http://dx.doi.org/10.5772/intechopen.70746
Authored by Masaaki Suzuki, Hiroko Koyama, Hideki Ishii, Koichi Kato, Margareta Ögren and Hisashi Doi

Contributors
Masaaki Suzuki, Hideki Ishii, Koichi Kato, Margareta Ögren, Hiroko Koyama, Hisashi Doi

Published by InTech
Janeza Trdine 9, 51000 Rijeka, Croatia

Publishing Process Manager Viktorija Zgela
Technical Editor SPi Global
Cover InTech Design team

First published February, 2018
Printed in Croatia
Legal deposit, Croatia: National and University Library in Zagreb

Additional hard copies can be obtained from orders@intechopen.com

Green Process of Three-Component Prostaglandin Synthesis and Rapid 11 C Labelings for Short-Lived PET Tracers,
Authored by Masaaki Suzuki, Hiroko Koyama, Hideki Ishii, Koichi Kato, Margareta Ögren and Hisashi Doi
p. cm.
Print ISBN 978-953-51-3764-1
Online ISBN 978-953-51-3765-8

For EU product safety concerns:
IN TECH d.o.o., Prolaz Marije Krucifikse Kozulić 3, 51000 Rijeka, Croatia,
info@intechopen.com or visit our website at intechopen.com.

Contents

Preface

This book is devoted to highly polished and potentiated novel C couplings developed by our group (M. Suzuki) in the course of a long-term, tight collaboration of chemistry, biology, and medicine experts, focusing on advanced organometallic techniques for the construction of a whole prostaglandin (PG) framework and new methodology for introducing short-lived ^{11}C radionuclide into carbon frameworks. Organic synthesis is now harmonized with other scientific fields and together they are recognized as commonly significant subjects. Truly efficient synthesis elaborated from practical and environmental points of view is demanded. The radiochemical probe for in vivo study should be designed with regard to the safety of human and natural environments. The first half of the book starts from our one-pot three-component coupling PG synthesis by combining organocopper or organozincate conjugate addition of an ω-side chain with 4-hydroxy-2-cyclopentenone, followed by controlled monoalkylation with an α-side chain, giving a common intermediate for general synthesis of PGs. Stable prostacyclin analogues, isocarbacycin, 15*S* -APNIC, a photoaffinity labeling probe for the prostacyclin receptor, and 15*R*-TIC selectively binding to a novel prostacyclin receptor in the central nervous system are demonstrated as further applications. The book ends with successful realization of the ultimate green coupling process. The second half of the book deals with rapid *C*-[^{11}C]methylation using Pd^0-mediated rapid cross-coupling between [^{11}C]methyl iodide and an organotributylstannane or organoboronate, and likely rapid *C*-[^{18}F]fluoromethylation using [^{18}F]fluoromethyl bromide and organoboronate. Rapid *C*-[^{11}C]carbonylation is also briefly described. ^{11}C-labeling methods are applied to syntheses of various biologically significant and disease-oriented positron emission tomography probes for in vivo imaging aiming at drug discovery and diagnosis. The book is written to attract the interest of scientists working in chemistry and interdisciplinary research areas such as medicinal chemistry, radiochemistry, chemical biology, molecular imaging, and related industries.

Masaaki Suzuki,
National Center for Geriatrics and Gerontology,
Japan

Chapter 1

Green Process of Three-Component Prostaglandin Synthesis and Rapid ^{11}C Labelings for Short-Lived PET Tracers: Highly Polished *C*-Couplings Revolutionizing Advances in Bio- and Medical Sciences

Masaaki Suzuki, Hiroko Koyama, Hideki Ishii,
Koichi Kato, Margareta Ögren and Hisashi Doi

Additional information is available at the end of the chapter

http://dx.doi.org/10.5772/intechopen.72868

Abstract

General synthesis of prostaglandins (PGs) has been accomplished based on a one-pot three-component coupling using a combination of organocopper or organozincate conjugate addition to 4-hydroxy-2-cyclopentenone followed by trapping of resulting enolate with an organic halide. Based on the use of this synthetic methodology, biologically significant PG derivatives including ent-Δ^7-PGA_1, 15*S*APNIC ([^{3}H]APNIC), and 15*R*–TIC have also been synthesized. Ultimately, organozincate conjugate addition combined with the enolate trapping by an organic triflate results in practical green three-component coupling comprising the use of stoichiometric amounts of three components (enone, α- and ω-side chains in a nearly 1:1:1 ratio) without using HMPA and heavy metals. General methodology for introducing short-lived ^{11}C and ^{18}F radionuclides into carbon frameworks has been established by developing rapid *C*-[^{11}C]methylation and *C*-[^{18}F]fluoromethylation using Pd^0-mediated rapid cross-coupling between [^{11}C]methyl iodide and an organotributylstannane or organoboronate; or [^{18}F]fluoromethyl bromide and organoboronate, respectively, allowing the synthesis of a wide variety of biologically significant and disease-oriented PET probes such as 15*R*-[^{11}C]TIC. Moreover, Pd^{II}-mediated rapid *C*-[^{11}C]carbonylation using [^{11}C]CO and organoboronate at ambient temperature under atmospheric pressure using conventional helium carrier gas has been explored. Further, *C*-[^{11}C]carboxylation has been promoted using [^{11}C]CO_2 and organoboronate with Rh^I catalyst under atmospheric pressure.

Keywords: *C*-couplings, three-component PG synthesis, synthesis of biologically significant PG derivatives such as ent-Δ^7-PGA_1, 15*S*–APNIC, and 15*R*–TIC, green practical three-component coupling, general rapid ^{11}C- and ^{18}F-labelings, short-lived PET probes, rapid *C*-[^{11}C]methylations, rapid *C*-[^{18}F]fluoromethylation, rapid *C*-[^{11}C] carbonylation, rapid *C*-[^{11}C]carboxylation, human and environmental sciences

1. Introduction

Organic synthesis has played a pivotal role so far in life science mainly by: (1) supplying enough amounts of significantly biologically active molecules difficult to obtain from natural sources; (2) enhancing the selectivity of biological activities as well as increasing chemical and biological stabilities by the structural modification based on rational molecular designing. In recent years, new research areas of "chemical biology" and "molecular imaging" have become increasingly important, particularly by adding new trends in drug discovery and in imaging diagnostics technologies. Accordingly, organic synthesis, with its huge potential in harmony with such advancing interdisciplinary scientific areas are strongly demanded, particularly for: (1) creating specific molecular probes to identify a target molecule (receptor) involved in the mechanism of a biological activity; (2) and clarifying the behavior of molecules in *in vivo* systems in terms of verifying drug efficacy and safety. In order to promote non-invasive dynamic *in vivo* molecular science such as a PET study, many efforts to accelerate the rate of the reaction are needed for labeling organic molecules with short-lived radioisotopes. Described herein are highly polished and potentiated novel C-couplings developed by our group in the course of long-term tight collaboration of chemistry, biology, and medicine [1–5].

2. Prostaglandin synthesis made easy by three-component coupling

Prostaglandins (PGs) exhibit diverse biological activities controlling a wide range of physiological functions in the circulatory, respiratory, and digestive systems and are also involved in vital defense processes such as inflammation, tissue repair, and immune response [6]. Although such activities have attracted much attention in view of high potentials as therapeutic agents, the supply of PGs from natural sources based on biosynthesis via PG-endoperoxide starting from arachidonic acid (**Scheme 1**) is difficult [7]. Among chemical syntheses more than 450 reported so far, Corey's synthesis [7] is a great triumph in synthetic organic chemistry, allowing not only commercial production but also contributing enormously to the progress of life science. The process based on linear synthesis requires 17 steps and additional optical resolution for the synthesis of PGF$_2\alpha$. The way to other PGs needs further additional steps. Obviously, "the three-component coupling process," namely the combination of three units, five-membered ring, α-chain, and ω-chain to lead a whole PG framework, would be an ideal approach in view of directness and synthetic flexibility (**Scheme 2**) [8]. Concretely, the three-component coupling comprising the conjugate addition of metalated ω-side chain unit to the protected 4-hydroxy 2-cyclopentanone followed by electrophilic trapping of the resulting enolate intermediate by an

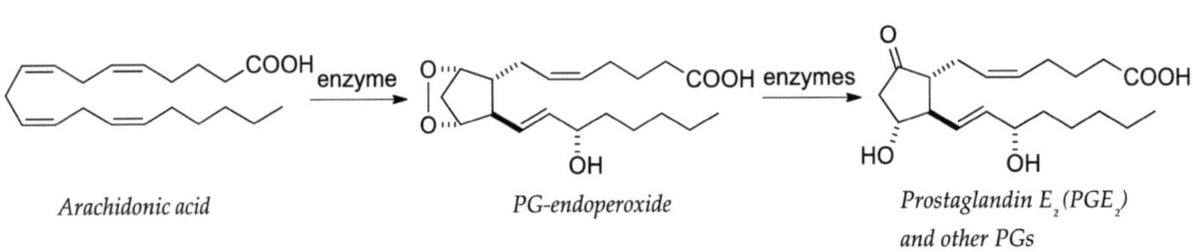

Scheme 1. Biosynthesis of prostaglandins.

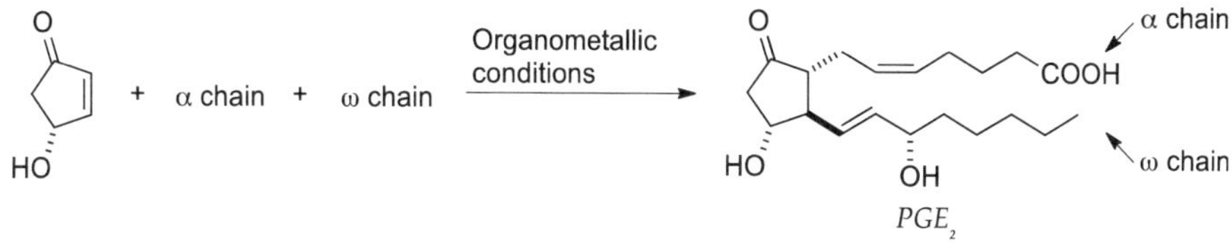

Scheme 2. Chemical synthesis (three-component synthesis, a long-lasting problem in synthetic chemistry).

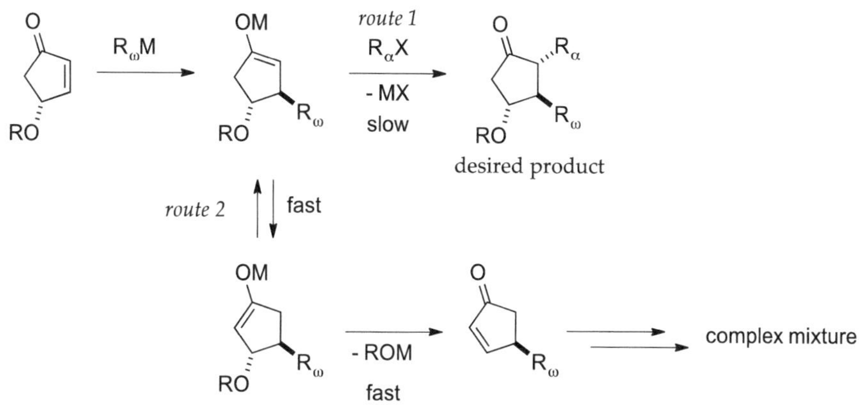

Figure 1. Problems of the three-component prostaglandin synthesis.

α-side chain of organic halide (**Figure 1**, *route* 1) had been planned for this purpose. However, this convergent approach had been a long-lasting problem until our success (see the following section) bcause the difficulty of controlling the enolate reactivity, namely the enolate, undergoes the facile proton exchange during alkylation inducing dehydration as a major side reaction, and eventually, decomposing to form a complex mixture (**Figure 1**, *route* 2) [8].

2.1. Highly selective mono-methylation of cyclopentanone lithium enolate with methyl iodide

Cyclopentanone lithium enolate (**1**) tends to undergo polyalkylation in THF without additives (**Table 1**, Entry 1), and therefore, this enolate was chosen as a model for realizing controlled mono-alkylation with an alkyl halide (**Table 1**). Thus, clean mono-alkylation of the lithium enolate with methyl iodide has been established in the combination of $(C_6H_5)_3SnCl$ (1 equiv) and HMPA (5 equiv) as additives at −30°C for 6 h to give 1-methylcyclopentanone with 77% yield (Entry 3) [9]. The combination of $(CH_3)_2Zn$ (1 equiv) and HMPA as additive is also efficient; the reaction at −78°C for 20 h, gives the same product with 95% yield (Entry 4) [10]. Here, HMPA accelerates the reaction (Entry 2) [10, 11] and $(C_6H_5)_3SnCl$ or $(CH_3)_2Zn$ suppresses the polyalkylation. Notably, it was proven that the reactive species is HMPA-coordinated lithium enolate and that LiI is not involved in the reaction.

2.2. Extremely short-step synthesis of PGE_2 by three-component coupling

The conditions thus obtained have successively being applied to the synthesis of PGE_2 (**6**) (**Scheme 3**), realizing the three-component coupling synthesis of PGs (via direct alkylation

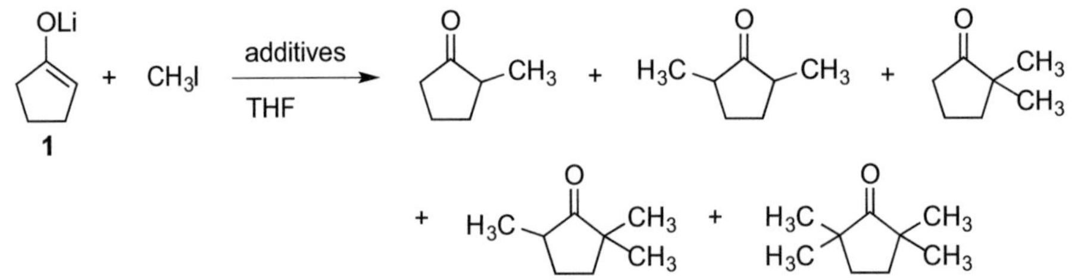

Entry	CH_3X (equiv)	Additives (equiv)	Temp., °C	Time, h	Yield, %	
					Monoalkyl ketone	Polyalkyl ketone
1	CH_3I (5)	—	0	1	51	33
2	CH_3I (5)	HMPA (5)	−78	20	**95**	3
3	CH_3I (5)	$(C_6H_5)_3SnCl$ (1), HMPA (5)	−30	6	**77**	0
4	CH_3I (5)	$(CH_3)_2Zn$ (1), HMPA (5)	−78	20	**95**	0
5	CH_3OTf (1.2)	—	−78	1	98	2
6	CH_3OTf (1.2)	$(CH_3)_2Zn$ (1)	−78	2	**100**	0
7	CH_3OTf (1.2)	$(CH_3)_2Zn$ (3)	−78	2	**100**	0

Table 1. Highly selective mono-methylation of cyclopentanone lithium enolate with methyl iodide.

route) for the first time. Here, the conjugate addition was performed by using the newly devised organocopper reagent [12] and the protected (*R*)-4-hydroxy-2-cyclopentanone (**2**) at 1:1 ratio in order to avoid the presence of extra nucleophiles in the reaction system. The combination of this conjugate addition and the trapping of the resulting metalated enolate with an excess amount of α-side chain iodide (**4**) in the presence of HMPA and $(C_6H_5)_3SnCl$ led to the successful construction of a whole PG skeleton [9, 13]. Thus, the metalated ω-side chain (**3**) was prepared in situ by treating the corresponding ω-side chain iodide with *tert*-C_4H_9Li (2 equiv) followed by the addition of CuI-$(n\text{-}C_4H_9)_3P$ (1:2.6 equiv ratio) at −78°C in THF. After conjugate addition to (**2**) at −78°C in THF, the resulting metalated enolate was quenched by α-side chain allyl iodide (**4**) (5 equiv) in the presence of HMPA (5 equiv) and $(C_6H_5)_3SnCl$ (1 equiv) at −30 to −20°C for 39 h in THF to give the desired coupling product (**5**) with 78% yield. Subsequent deprotection followed by ester hydrolysis afforded PGE_2 (**6**) at 83% yield.

2.3. General synthesis of prostaglandins

The alkylation of the enolate intermediate generated by the conjugate addition using a stoichiometric amount of the organocopper phosphine complex [12] with (**2**) at −78°C in THF with methyl 7-iodo-5-heptynylate (**7**, 5 equiv) in the presence of HMPA (11 equiv) and $(C_6H_5)_3SnCl$ (1 equiv) in THF at −30°C for 39 h gives 5,6-dehydro-PGE_2 methyl ester (**8**) at 76% yield (**Scheme 4**, Conditions A) [13]. It was found that the conjugate addition proceeds by using triorganozincate species generated by a 1:1 mixing of ω-side chain lithium and $(CH_3)_2Zn$

http://dx.doi.org/10.5772/intechopen.72868

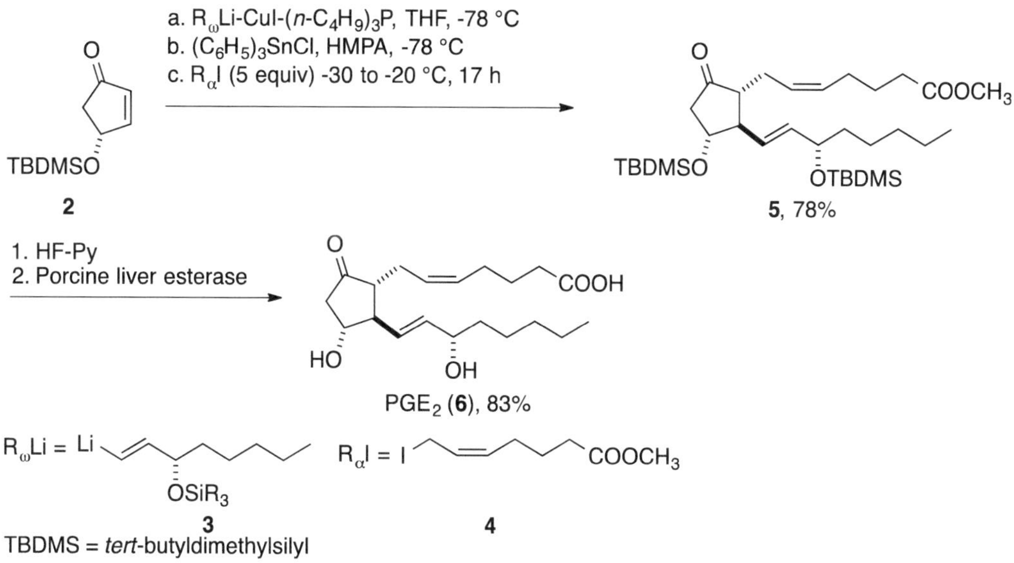

Scheme 3. Extremely short-step synthesis of prostaglandin E_2 (**6**) by three-component coupling.

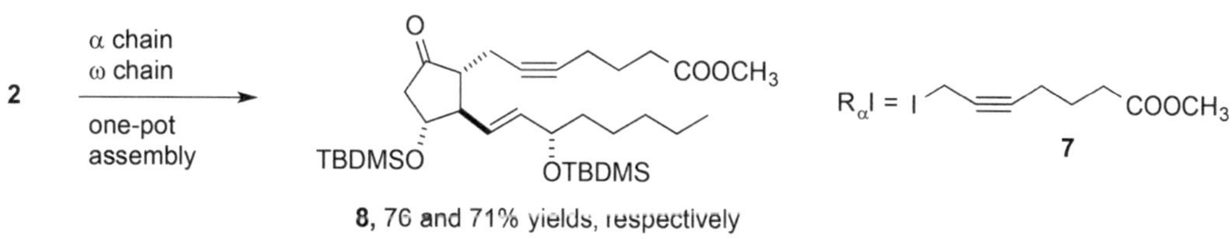

Conditions A: a. R_ωLi-CuI-2,6-(n-C_4H_9)$_3$P, THF, -78 °C; b. $(C_6H_5)_3$SnCl (1 equiv), HMPA (11 equiv), -78 °C; c. R_αI (**7**, 5 equiv), -30 °C, 39 h, 76%
Conditions B: a. R_ωLi-$(CH_3)_2$Zn, THF, -78 °C; b. R_αI (**7**, 5 equiv), HMPA (10 equiv), -78 to -40 °C, 24 h, 71%

Scheme 4. Synthesis of 5,6-dehydro PGE_2 methyl ester (**8**).

without a transition metal [10, 14]. Similarly, the coupling of the enolate thus prepared by the conjugate addition with propargylic iodide (**7**, 5 equiv) in the presence of HMPA (10 equiv) in THF at −78 to −40°C for 24 h, after workup and chromatographic separation, gives (**8**) at 71% yield (**Scheme 4**, Conditions B). The isolation of the product in latter process is much easier than the former. General syntheses of PGs have been accomplished via the propargylic intermediate (**8**) in several steps including selective hydrogenation leading to PGE_1, PGE_2, $PGF_1\alpha$, and $PGF_2\alpha$ in addition to Pd^{II}-mediated intramolecular cyclization of a hydroxyl group at the C(9) to the C(7) positions, respectively, followed by depalladation with ammonium formate to generate prostacyclin (PGI_2) in an extremely short way (**Figure 2**) [14, 15].

The enolate species in situ prepared by the conjugate addition of an ω-side chain to (**2**) can also be trapped by other electrophiles such as α-chain aldehyde [16] and nitro-olefin, [17] to synthesize PGs and biologically significant PG analogs such as Δ^7-PGA_1 (**9**), [18] ent-Δ^7-PGA_1 (**10**), [19] and 6-oxo-PGE_1 (**11**) [17].

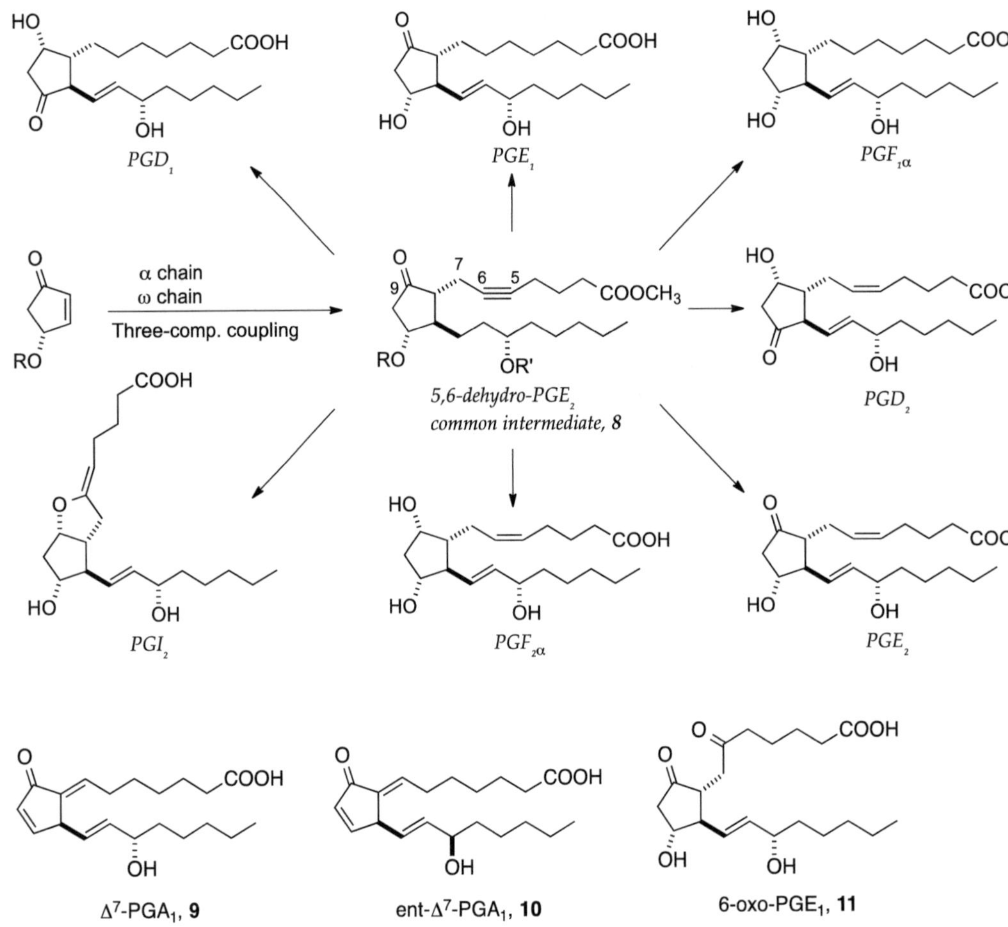

Figure 2. General syntheses of prostaglandins.

2.4. Synthesis of isocarbacyclin

Isocarbacyclin (**12**), a stable analog of the chemically unstable PGI_2, created synthetically by Ikegami [20] has potent PGI_2 like activities. The structure is featured by the olefinic bond at C (6) and C(9α) positions in the bicyclo[3,3,0]octane skeleton. Our synthetic plan to introducing this structure stems on radical cyclization between carbons at the C(9α) and C(6) positions by 5-exo-dig manner and subsequent γ-protodesilylation of an allylsilane (**Scheme 5**) [21] The accomplished synthetic route is illustrated in **Scheme 6**. Thus, starting from the acetylenic intermediate (**8**), sequencing reactions of methylenation of the carbonyl at C(9) position gives (**13**) stereoselective hydroboration, followed by H_2O_2 oxidation giving (**14**), pyridinium dichromate (PDC) oxidation, and silylation giving the silyl alcohol (**15**). After conversion to *m*-(trifluoromethyl) benzoate, photolysis gave the allylsilane (**16**) with 1:1 stereoisomeric mixtures, and then treated with aqueous $HClO_4$ for deprotection and CF_3COOH for protodesilylation to give isocarbacyclin methyl ester (**17**) regiospecifically, which undergoes hydrolysis facilely to afford isocarbacyclin (**12**).

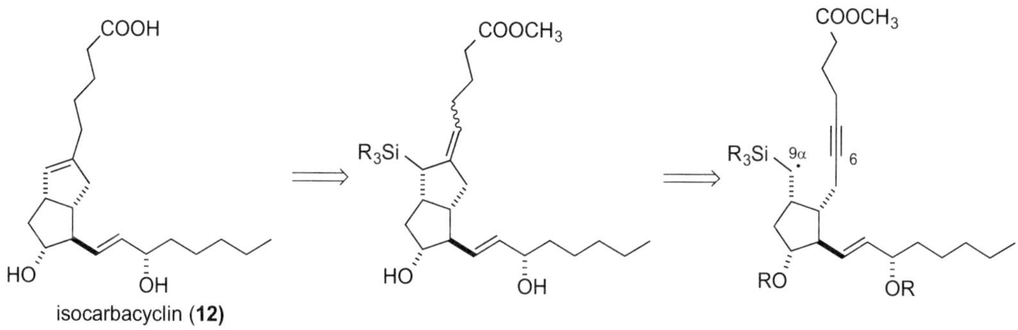

Scheme 5. Strategy for regiospecific construction of cyclopentenone structure in isocarbacyclin (12).

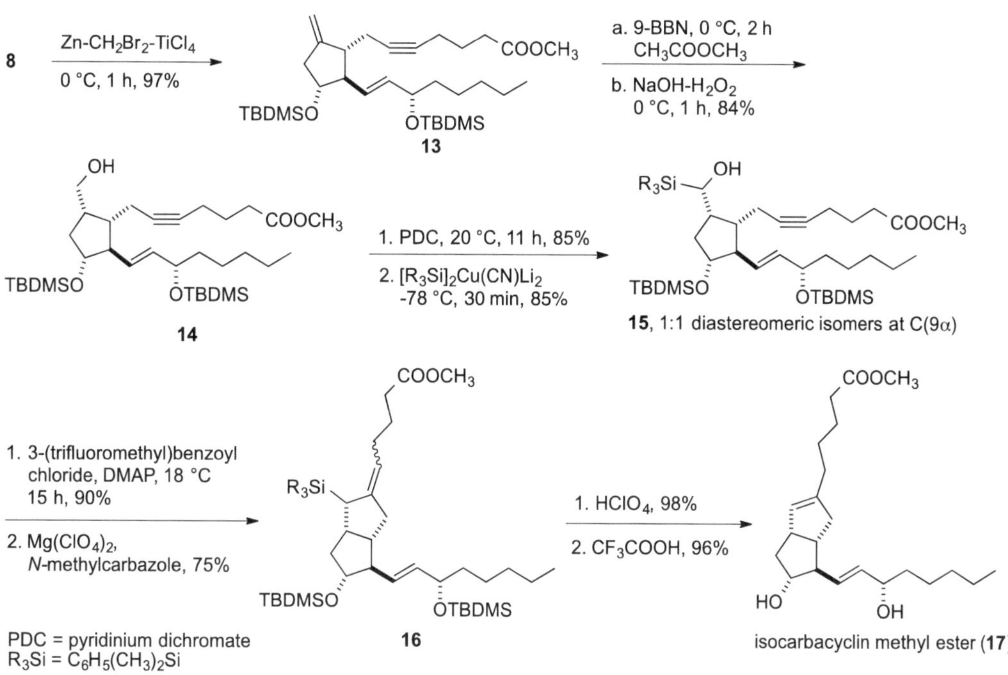

Scheme 6. Synthesis of isocarbacyclin methyl ester (17).

2.4.1. *Synthesis of azide-functionalized isocarbacyclin analog, 15S–APNIC and radio-labeled photoaffinity probe, 15S-[^{3}H]APNIC for identification of a prostacyclin receptor (IP)*

Prostacyclin (PGI_2) is particularly a potent vasodilator and an inhibitor of platelets aggregation among PGs. It activates adenylate cyclase in platelets, vascular smooth muscles, NCB-20 cells, and mastocytoma-p-815 cells. However, there has been little progress in the study of the structure of the PGI_2 receptor protein (IP) because of its low concentration in cell membranes and the lack of a suitable antagonist making it difficult to solubilize the receptor protein to a homogeneous state without loss of the binding activity. Here, we have been intrigued by the use of a photoaffinity labeling method as another tool for the identification of the receptor

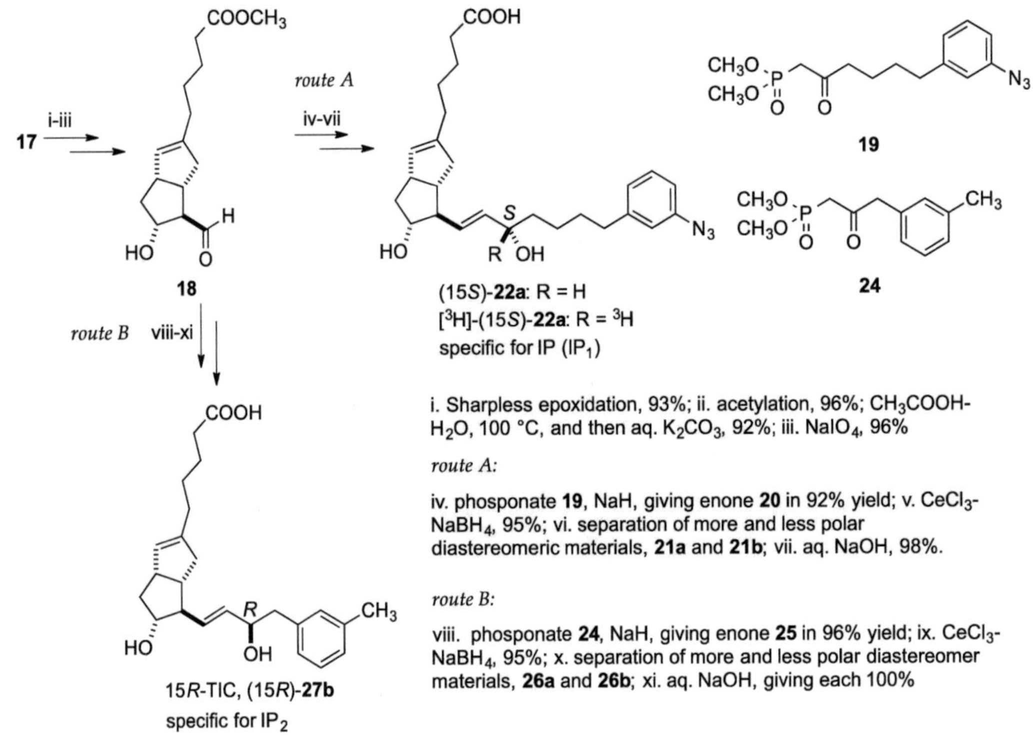

Scheme 7. Synthesis of stable prostaglandin I_2 analogs, 15*S*-APNIC ((15*S*)-**22a**) and 15*R*-TIC ((15*R*)-**27b**).

protein. A novel molecular probe for this purpose has been designed and synthesized as follows (**Scheme 7**, *route A*) [22]: the ω-side chain in isocarbacyclin methyl ester (**17**) was readily degraded by selective epoxidation of an allylic olefin with Sharpless method followed by the epoxide ring opening and subsequent $NaIO_4$ oxidation to give the aldehyde (**18**), which further underwent the carbon elongation to reconstruct the PG skeleton by Horner-Emmons type condensation with the azide-functionalized phosphonate (**19**), resulting in affording the enone (**20**). The reduction of the carbonyl in (**20**) using $NaBH_4$ in the presence of $CeCl_3$ gives allylic alcohol in a 1:1 diastereomeric mixture, which can be separated by silica gel chromatography to afford pure esters of azide-functionalized isocarbacyclin analogs, APNIC methyl esters (**21a**) and (**21b**) as more and less polar materials, respectively. Here, *S*- and *R*-configurations at the C(15) position of these diastereoisomers were determined by chemical transformations from methyl ester of APNIC (15*S*-APNIC methyl ester) and defined optically active by (*S*)-glycerol 1,2-acetonide, respectively to give the same (*S*)-2-hydroxy-6-phenylhexanol [22]. Finally, each ester underwent alkali hydrolysis, giving (15*S*)-**22a** and (15*R*)-**22b**, respectively. The tritium labeled 15*S*-APNIC ([^{3}H]-(15*S*)-**22a**) is prepared commercially (Amersham International in England) under the same reduction conditions using [^{3}H]$NaBH_4$ with $CeCl_3$, followed by alkali hydrolysis.

The aldehyde intermediate (**18**) can also be prepared by deprotection (CH_3COOH/H_2O/THF) of the corresponding THP-protected aldehyde compound [23] derived from Corey lactone [7].

Results of binding assay [22, 24]: (15*S*)-**22a** has the highest affinity among the azidophenyl derivatives and exhibited an IC_{50} value of 3 nM (**Figure 3**).

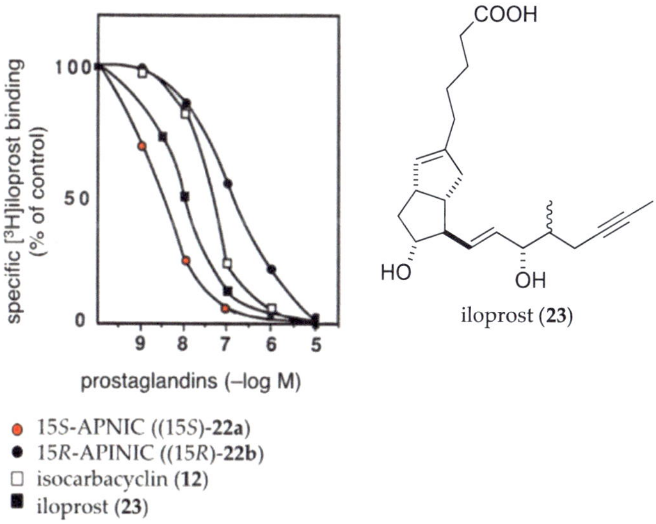

Figure 3. Binding affinity with a PGI_2 receptor: Displacement of [^{3}H]iloprost by azidophenyl derivatives ((15*S*)-**22a**, (15*R*)-**22b**) vs. iloprost (**23**).

Biological activity (**Figure 4**) [22, 24]: the activity study of (15*S*)-**22a** as an agonist on the stimulation of adenylate cyclase in the mastocytoma P-815 membrane fraction using iloprost (**23**) as a standard compound exhibited that the half-maximal concentration to stimulation was 50 nM, convincing us that (15*S*)-**22a** has considerable agonist character for the PGI_2 receptor (IP).

Identification of IP (**Figure 5**) [24]: photoaffinity labeling of the PGI_2 receptor was conducted by UV irradiation using the membranes of porcine platelets with [^{3}H]-(15*S*)-**22a**, followed by SDS-PAGE to give the results shown in **Figure 5**—the left panel shows radioactivity in the gel slice; the right, fluorography, lane1 ([^{3}H]-(15*S*)-**22a** only), lane 2 (in the presence of GTPγS), lane 3 (in the presence of (**23**)). Thus, photoaffinity labeling entraps a 45 kDa protein in porcine platelets membranes. The protein is also sensitive to GTPγS and the binding [^{3}H]-(15*S*)-**22a** is blocked by iloprost.

2.4.1.1. Synthesis of tolyl-functionalized isocarbacyclin analog, 15R–TIC, a ligand specifically binding with novel prostacyclin receptor subtype (IP_2) in the central nervous system

Several lines of evidence suggest that PGI_2 also has neuromodulatory actions. A quantitative *in vitro* autoradiographic mapping of the PGI_2 receptor in the brain using a stable specific agonist for a PGI_2 receptor demonstrated a high density distribution in four regions of the lower brain stem including the medial and commissural subnuclei of the nucleous tractus solitarius (NTS). Precise analysis of the role of PGI_2 in the brain requires the development of a specific molecular probe capable of sharply discriminating the actions in the central and peripheral nervous systems, respectively. In this context, our interest has been directed toward designing a new ligand selectively responsive to a receptor in the central nervous system [1]. Thus, the title compound 15*R*-TIC specifically binding with a novel PGI_2 receptor in the central nervous system has been devised as follows [25]: the steps are similar to the synthetic scheme of APNIC as shown in *route B* (**Scheme 7**); the Horner-Emmons type condensation of (**18**) with

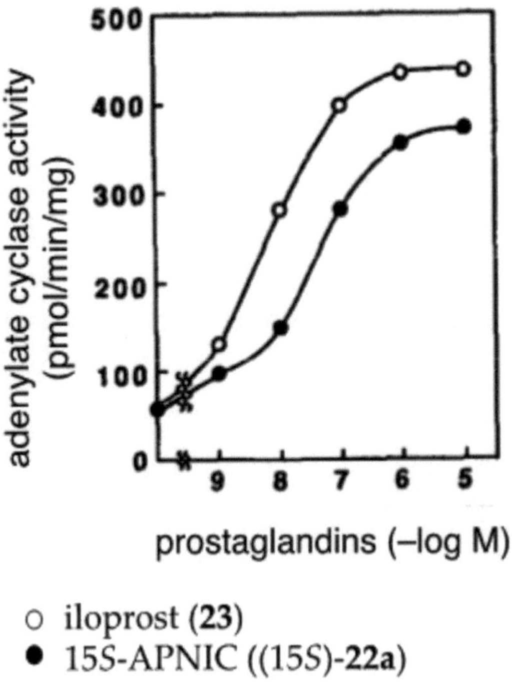

Figure 4. Effect of iloprost (**23**) and 15*S*–APNIC ((15*S*)-**22a**) on adenylate cyclase activity in the presence of GTP.

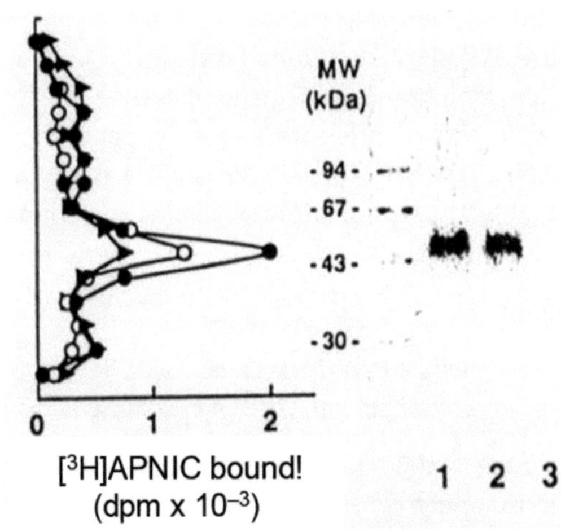

Figure 5. Photoaffinity labeling of the membrane of porcine platelets with [^{3}H]APNIC. A 45-kDa protein was labeled by photolysis in the presence of [^{3}H]APNIC ([^{3}H]-(15S)-**22a**). Porcine platelet membranes were incubated with [^{3}H]APNIC in the absence (lane 1) or presence of GTPγS (lane 2) or iloprost (**23**) (lane 3).

the phosphate (**24**) to elongate the carbon chain gives the enone (**25**) and subsequent reduction in the presence of $CeCl_3$ gives the diol with a 1:1 mixture of diastereomers at the C(15) position. The stereoisomers are separated by silica gel chromatography to give stereochemically pure alcohols, **26a** and **26b**, as different polarities on TLC (R_f = 0.18 and 0.26 with 1:1 ethyl acetate and hexane as solvent), respectively. Finally, alkali hydrolysis of **26a** and **26b** affords the corresponding tolyl-functionalized isocarbacyclin analogs, 15*S*–TIC ((15*S*)-**27a**) and 15*R*–TIC ((15*R*)-**27b**), respectively. The configuration of the carbon center at the C(15) position is determined by chemical transformations similar to that in the synthesis of APNIC starting from (**26a**) and optically active (*S*)-glycerol 1,2-acetonide (**30**) to give the same tolyl-containing bis (MTPA)ester (**29**) (**Scheme 8**).

The binding assay for PGI_2 receptor proteins in frozen tissue sections containing either the thalamus or the NTS as representative of the central or peripheral nervous systems, respectively, was conducted by the replacement of 10 nM tritium-labeled isocarbacyclin ([^{3}H] isocarbacyclin) with nonradioactive derivatives as shown in **Figure 6** [26]. It was discovered that 15*R*–TIC ((15*R*)-**27b**), which exhibited the highest binding affinity for the receptor in the thalamus (IC_{50} = 31 nM) among other TIC derivatives, showed very weak binding with the receptor in NTS (IC_{50} = 1 μM). The 15*S* isomer (15*S*)-**27a** showed slightly lower binding (IC_{50} = 38 nM) in the thalamus than (15*R*)-**27b** and maintains considerably strong binding affinity (IC_{50} = 32 nM) in the NTS. Other TIC derivatives showed weaker thalamus binding affinity. Isocarbacyclin (**12**), a lead compound for the study, exhibited high binding affinity both in the thalamus (IC_{50} = 31 nM) and the NTS (IC_{50} = 23 nM). The binding profile of 15*R*–TIC ((15*R*)-**27b**) distinctly contrasted with that of cicaprost (**32**), a stable PGI_2 receptor agonist in the iloprost family, which showed high binding and selectivity for the receptor in the NTS (IC_{50} = 30 nM). Consistent with the binding property, 15*R*–TIC showed a very weak inhibitory effect on platelets aggregation (IC_{50} > 400 nM), while cicaprost (**32**) and isocarbacyclin (**12**) exhibited potent inhibitory effects (IC_{50} = 3.2 and 2.5 nM, respectively). Thus, we succeeded in

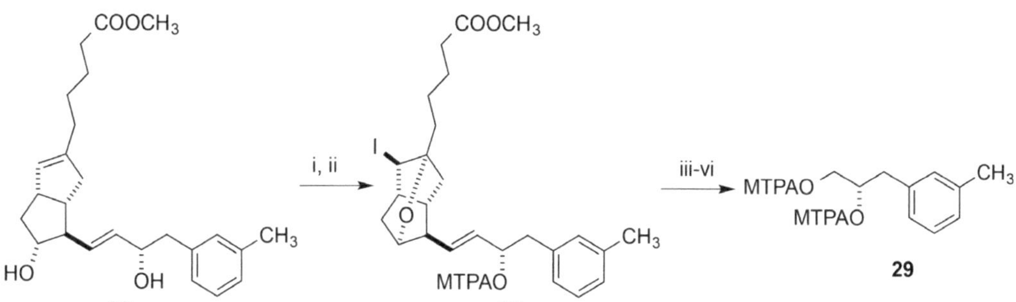

26a **28**

i. I_2, CH_3OH/H_2O (3:1); ii. (*S*)-α-methoxy-α-(trifluoromethyl)phenylacetyl chloride [(*S*)-MTPACl], pyridine; iii. OsO_4, pyridine; iv. $NaIO_4$, THF/ether/H_2O; v. $NaBH_4$, CH_3OH; vi. (*S*)-MTPACl, pyridine

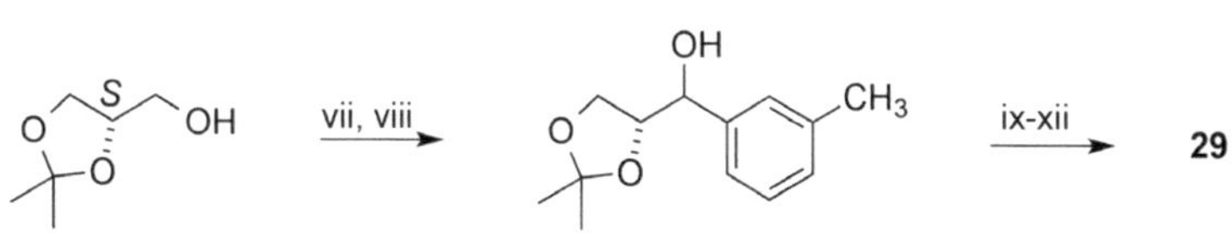

30 **31**

vii. PCC, CH_2Cl_2; viii. *m*-$CH_3C_6H_4MgCl$, ether; ix. NaH, and then CS_2, CH_3I; x. (*n*-C_4H_9)$_3$SnH; xi. CF_3COOH, THF, H_2O; xii. (*S*)-MTPACl, pyridine

Scheme 8. Determination of the absolute configuration at C(15) of the isocarbacyclin analog by chemical correlation.

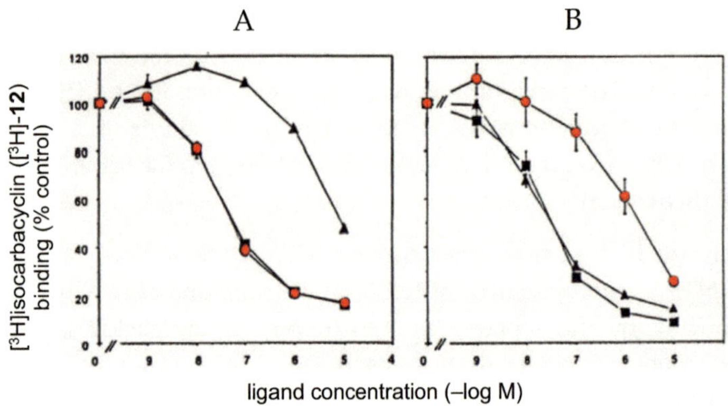

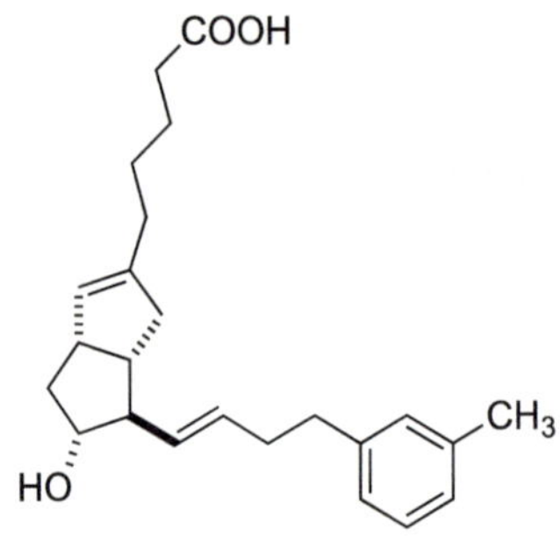

15-deoxy-TIC, **33**

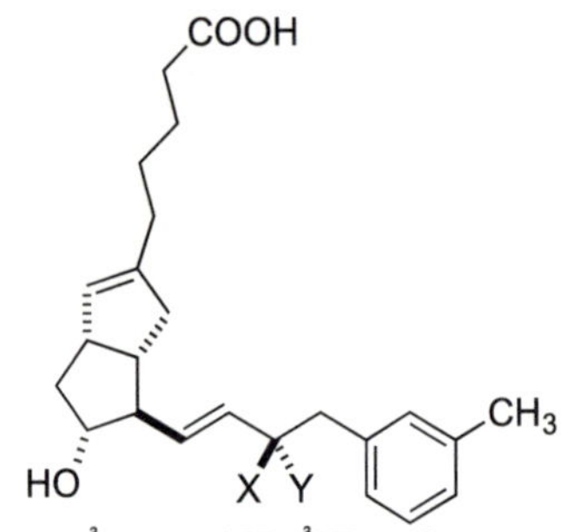

X = ^{3}H, Y = OH: [^{3}H]-(15*S*)-**27a**
X = OH, Y = ^{3}H,: [^{3}H]-(15*R*)-**27b**

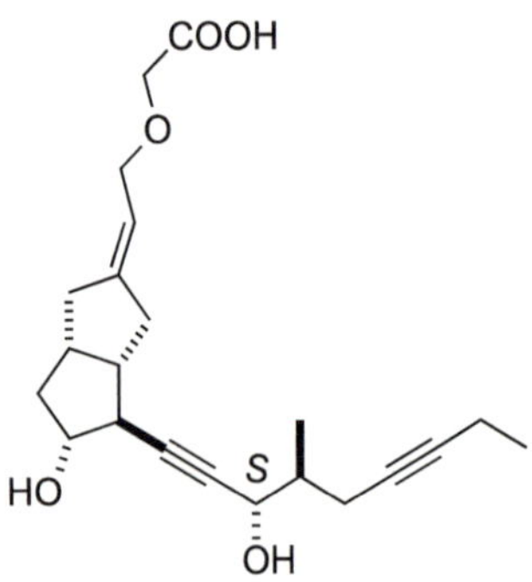

cicaprost (**32**)

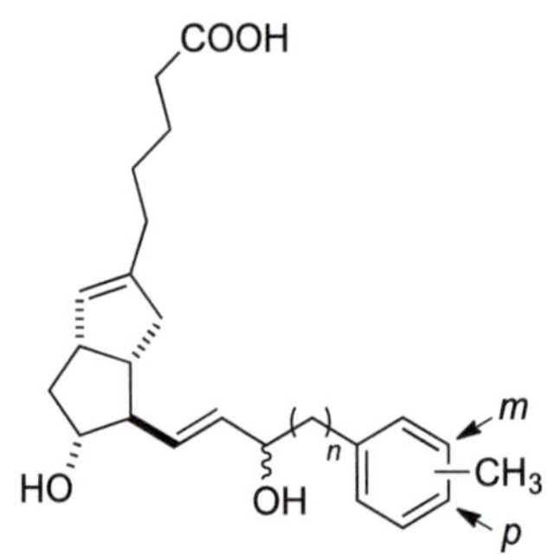

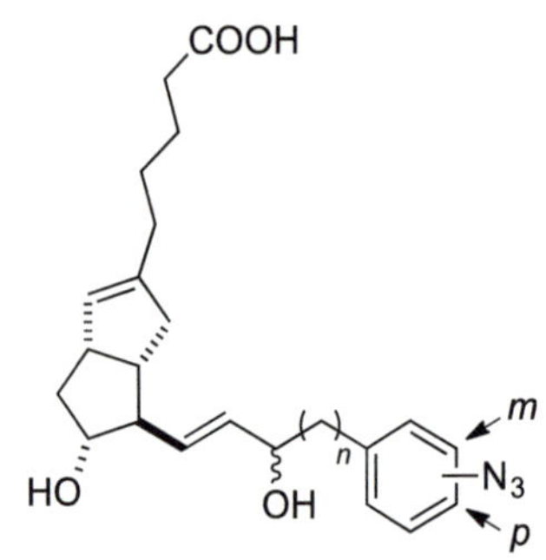

Figure 6. Binding studies of 15R–TIC ((15R)-**27b**) for PGI_2 receptors in the central nervous system (IP_2) and the peripheral system (IP_1). Displacement of 10 nM [^{3}H]isocarbacyclin ([^{3}H]-**12**) binding by nonradioactive 15R–TIC ((15R)-**27b**, ●), isocarbacyclin (**12**, ■), and cicaprost (**32**, ▲) in thalamus (A) and the NTS (B) of frozen sections of rat brain A and thalamus B. NTS: Nucleus tractus solitaries.

finding a stable PGI_2 analog with high binding selectivity for the PGI_2 receptor (IP_2) in the central nervous system based on the structural modification of isocarbacyclin (**12**) [1, 25, 26]. In addition, later, 15-deoxy-TIC (**33**) with a binding potency for IP_2 ten times higher than 15*R*–TIC ((15*R*)-**27b**) and a weak inhibitory effect on platelet aggregation (IC_{50} > 400 nM) has been synthesized by the stepwise construction of the *E*-olefin structure in the ω-side chain [27]. Both

15*R*–TIC ((15*R*)-**27b**) and 15-deoxy TIC (**33**) showed distinct protective effect for hippocampal neuronal death under high oxygen (50%) atmosphere [28]. Such an anti-apoptotic effect was more potent than basic fibroblast growth factor (bFGF) at its maximum values. Here, (**33**) was about ten times more potent in the neuronal protection than (15*R*)-**27b** (IC_{50} = 30 and 300 nM, respectively), being in good correlation to their binding affinities for the IP_2 receptor [25, 26]. Furthermore, (15*R*)-**27b** strongly protected CA1 pyramidal neurons against ischemic damage in gerbils; thus, we can conclude that (15*R*)-**27b** acts as an effective neuronal survival-promoting factor both *in vitro* and *in vivo* vital systems [29]. To perform *in vitro* analyses of TICs, tritium-labeled TIC was needed. Thus, the tritium labeling of TICs, a mixture of 15*S*-[^{3}H] TIC ([^{3}H]-(15*S*)-**27a**) and 15*R*-[^{3}H]TIC ([^{3}H]-(15*R*)-**27b**), was conducted using commercial service (Amersham International in England) according to the same reduction scheme as described using [^{3}H]$NaBH_4$ with $CeCl_3$. It was then followed by alkali hydrolysis and reverse-phase HPLC separation into pure forms. The binding sites using [^{3}H]-(15*R*)-**27b** for a number of coronal sections of rat brain indicated that the binding was high in most of the thalamic regions, limbic structures, and some parts of the cortical regions [26]. The binding of [^{3}H]-(15*R*)-**27b** was displaced by 15*R*–TIC ((15*R*)-**27b**) efficiently (IC_{50} = ca. 10 nM)). The Scatchard analysis in rat brain sections using [^{3}H]-(15*R*)-**27b** showed that there were two binding sites with high and medium-high affinities for the IP_2 in the thalamus and striatum; the K_d values were 0.94 and 30 nM in the thalamus, and 0.54 and 37 nM in the striatum, respectively [26]. It was also found that the CNS-specific PGI_2 receptor (IP_2) was expressed mainly in neurons, but not in the presynaptic terminals of afferents or glial cells [26].

With regard to the influence of the length of the ω-side chain in TICs, we found that the binding affinity for the PNS-type PGI_2 receptor (so called IP; here IP_1) in the NTS increased in the order of: 16-*m* ($n = 1$) < 17-*p* ($n = 2$), 19-*p* ($n = 4$) < 17-*m* ($n = 2$) < 19-*m* ($n = 4$), while for the CNS-specific PGI_2 receptor (referred to as IP_2) in the thalamus, it increased in the order of: 19-*m*, 19-*p* ($n = 4$) < < 17-*p* ($n = 2$) < 17-*m* ($n = 2$) < 16-*m* ($n = 1$) [25, 26]. Thus, the receptor in the NTS favors longer side-chain derivatives, while the receptor in the thalamus matches more with shorter chain analogs. In both cases, the *meta*-methyl substituent on the aromatic ring was preferred. Overall, (15*R*)-**27b** exhibited the highest binding affinity for the CNS-specific PGI_2 receptor (IP_2) in the thalamus. It is also worth noting that the binding tendency obtained for the PNS-type PGI_2 receptor (IP_1) in the NTS is consistent with the binding mode of 15*S*–APNIC ((15*S*)-**22a**) for the PGI_2 receptor (IP) in mast cells and platelets as peripheral systems shown in the previous section [22]. The tolyl structure in (15*R*)-**27b** here was designed so that it could be used for *in vivo* molecular imaging study on the novel prostacyclin receptor subtype (IP_2) in the living brain after introduction of a short-lived ^{11}C radioisotope in the methyl group (see Section 3.1.2).

2.5. Ultimately clean three-component coupling using nearly stoichiometric amounts (1:1:1 ratio) of (*R*)-4-hydroxy-2-cyclopentanone, and α- and ω-side chains without using heavy metals and carcinogenic HMPA: Realization of practical green coupling process

The decreasing amounts of α-side chain unit and hazardous HMPA in our three-component coupling is important to realize an ideal stoichiometric green process (see **Scheme 4** and **Figure 2** in Sections 2.3 and 2.4). In this context, Gooding devised a two-step sequential process [30]: The trapping of the enolate species, generated by the conjugate addition of the ω-side chain vinyl lithium obtained from (**34**) in the presence of $Li_2(CH_3)_2Cu(CN)$ to get enone (**35**); reaction with trimethylsilyl chloride to synthesize the enol silyl ether (**36**); then the crude (**36**) is treated again with *n*-butyl lithium to generate pure lithium enolate, which undergoes alkylation with methyl 5,6-dehydro-7-triflyl-heptenoate ((**37**), 1.7 equiv) to give 5,6-dehydro-PGE_2 methyl ester (**38**) at 65% yield (**Scheme 9**). Based on this information, Lipshutz tried a one-pot synthesis of the whole skeleton of the PG by conjugate addition of the ω-side chain nucleophile generated by the treatment of the zirconylated ω-side chain unit in the presence of CH_3Li, $Li_2(CH_3)_2Cu(CN)$, and $(CH_3)_3ZnLi$ to give (**35**), followed by alkylation of the enolate intermediate with an excess amount of the same triflate (**37**, 6.0 equiv) to give 5,6-dehydro-PGE_2 methyl ester (**40**) as a stereoisomeric mixture (**15**) at 71% yield (**Scheme 10**) [31]. The need for an excess amount of α-chain triflate (**37**) to trap the enolate is presumably due to the complicated character of the enolate system including a transition metal after conjugate addition—there is still potential for further exploration. As shown in **Table 1** (continued) (see also **Table 1** in Section 2.1), interestingly, the trapping of the lithium enolate with methyl triflate in THF gave mono-methylation in 98% accompanied with 2% of polyalkylation product (Entry 5).

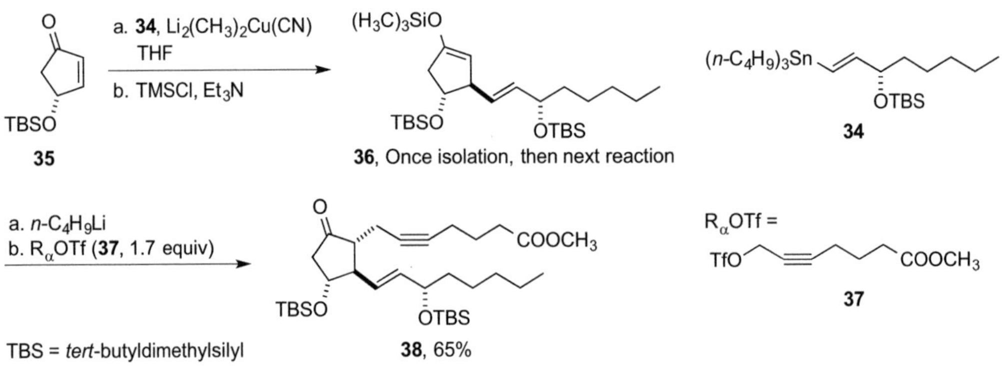

Scheme 9. Two-step protocol for the synthesis of a prostaglandin intermediate based on the alkylation of lithium enolate with propargylic triflate.

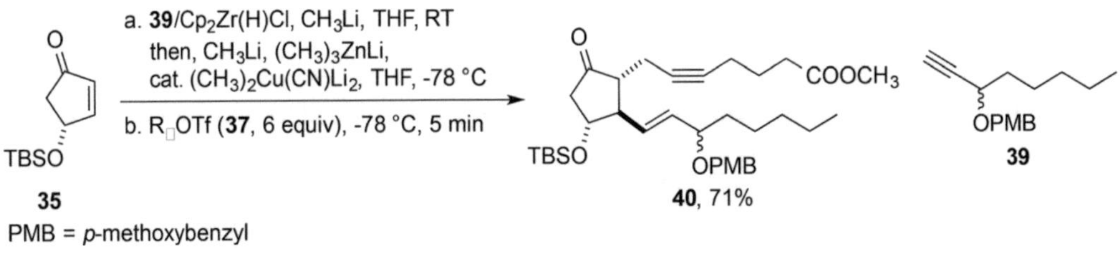

Scheme 10. One-pot three-component coupling based on the conjugate addition followed by enolate alkylation with an excess amount of propargylic triflate.

On the other hand, surprisingly, in the presence of dimethylzinc (1 to 3 equiv), the reaction gave *only* mono-methylation in quantitative yield as determined by GC analysis (Entries 6 and 7) [32]. These facts mean that the addition of dimethylzinc to the lithium enolate system completely suppresses the proton exchange causing polyalkylation; the reaction system still has enough reactivity for alkylation with methyl triflate even in the absence of HMPA. These fascinating findings reminded us of the combination with organozinc-aided conjugate addition previously developed by our group [10, 14] (see Section 2.3). Thus, a one-pot operation conducted by the conjugate addition of protected (*R*)-4-hydroxy 2-cyclopentanone (**2**) by ω-side chain vinyl lithium generated from vinyl stannane (**34**) (1 equiv) in the presence of 1 equiv of dimethylzincin THF at −85°C for 1.5 h followed by trapping of the resulting enolate with propargyl triflate (**37**) prepared by the corresponding alcohol (1.5 equiv for (**2**)) at −78°C for 2 h gave the protected 5,6-dehydro-PGE_2 methyl ester (**8**) at a high yield of 88% (**Scheme 11**, A) [32], which is a key intermediate for general synthesis of natural PGs (see **Figure 2** in Section 2.3) and isocarbacyclin (**12**) (see **Scheme 6**), whose methyl ester (**17**) can also lead to 15*S*–APNIC ((15*S*)-**22a**) and 15*R*–TIC ((15*R*)-**27b**) via the ω-side chain manipulation (see Sections 2.4.1.1 and 2.4.1.2). Most likely, the combination of the conjugate addition of ω-side chain lithium dimethylzincate generated by an ω-side chain lithium (generated from (**41**) with *n*-butyllithium in THF at −78°C, 1 h) and

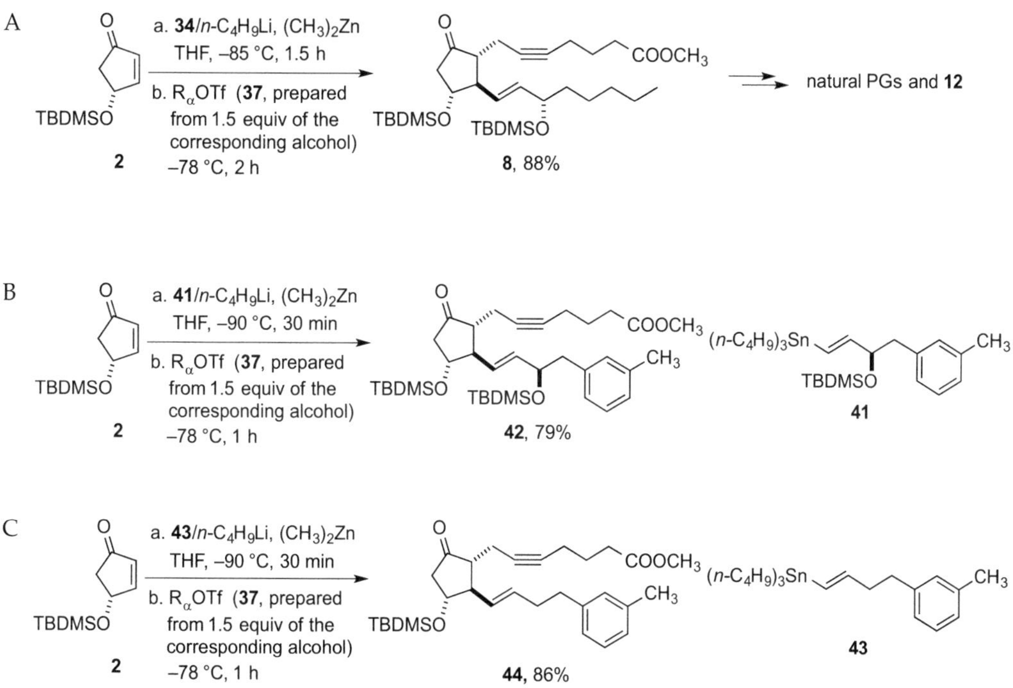

Scheme 11. Ultimate one-pot three component coupling comprising organozincate conjugate addition (1:1 ratio of ω-side chain and the enone) followed by alkylation of the resulting enolate with nearly stoichiometric amounts of the α-side chain propargyl triflate. A: three-component coupling using natural ω-side chain unit leading to a key intermediate for general synthesis of natural PGs and isocarbacyclin (**12**); B: three-component coupling using unnatural ω-side chain unit, potentially useful for the synthesis of 15*R*–TIC ((15*R*)-**27b**); C: three-component coupling potentially useful for the synthesis of 15-deoxy-TIC (**33**).

dimethylzinc (1 equiv) to the enone (**2**) at −90°C for 30 min followed by the trapping of the enolate with α-side chain triflate (**37**) (prepared from 1.5 equiv corresponding alcohol) at −78°C for 1 h gave the desired coupling product (**42**) at 79% yield (**Scheme 11**, B) [32]. Further, the combination of the conjugate addition of ω-side chain derived from (**43**) followed by trapping with (**37**) also gave (**44**) at 86% yield (**Scheme 11**, C) [32]. The latter two products, (**42**) and (**44**), are potentially key intermediates for the straightforward synthesis of 15*R*–TIC ((15*R*)-**27b**) and 15-deoxy-TIC (**33**), respectively (see Sections 2.4, 2.4.1.2, and 3.1.2). Thus, we succeeded in realizing an ultimate one-pot three-component coupling process called *green coupling process* based on a novel organozinc-aided conjugate addition/alkylation using stoichiometric amounts (nearly 1:1:1 ratio) of enone, and α- and ω-side chain units without using heavy metals and carcinogenic HMPA.

3. Efficient synthesis of short-lived ^{11}C- and ^{18}F-labeled PET probes by Pd^0-mediated rapid cross-coupling reactions, *C*-[^{11}C]methylation, *C*-[^{18}F]fluoromethylation, *C*-[^{11}C]carbonylation, and *C*-[^{11}C]carboxylation

Positron emission tomography (PET) is a noninvasive imaging technology with good resolution, high sensitivity, and accurate quantification, that assists in temporally and spatially analyzing the dynamic behavior of molecules in *in vivo* systems using a target-specific molecular probe labeled with positron-emitting radionuclides such as ^{11}C, ^{13}N, ^{18}F, and ^{76}Br (**Figure 7**) [33]. As shown in **Figure 7**, a positron (positively charged electron, e$^+$) emitted by a radionuclide collides with a nearby electron within a few millimeters inside the tissue to produce two high-energy γ-ray photons of 511 keV each. These photons travel in opposite directions (at 180°) in the body and can be detected by a pair of opposing scintillation

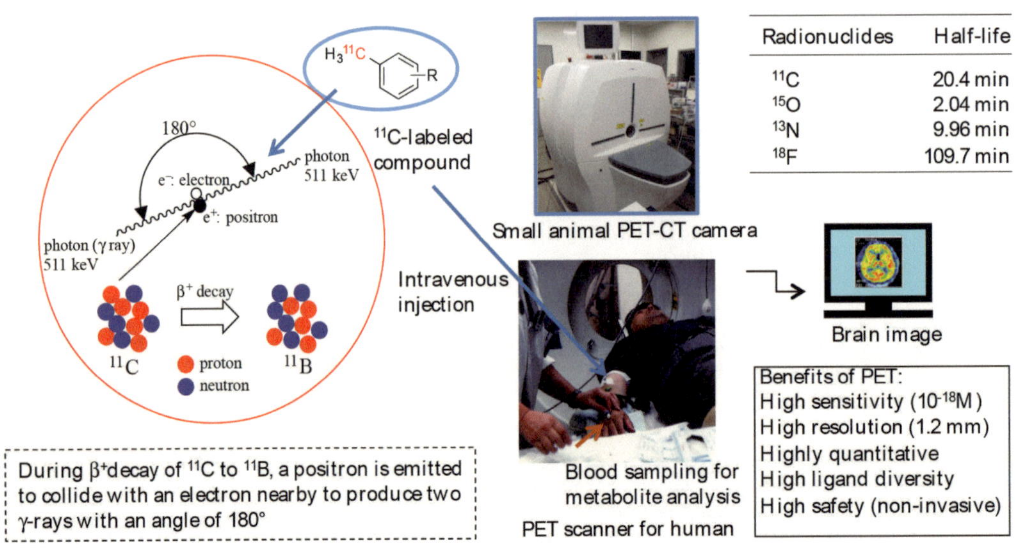

Figure 7. Principle of brain imaging by PET representatively shown by ^{11}C to ^{11}B decay.

detectors. If the two opposite detectors are hit by the positrons simultaneously, it indicates that the photons come from the same decay event. The data are then fed to a computer that reconstructs the three-dimensional tomographic image of the sample being analyzed. PET enables *in vivo* imaging using an extremely small mass of the compound (sub-femtomole), at extremely low concentrations (sub-picomolar) far below the critical concentration that would cause pharmacological effects. The advantages of PET technology include the following: (1) O, N, and C are ubiquitous elements in biologically active compounds occurring in nature; (2) high resolution and sensitivity; (3) a short half-life that makes it safe even with radiation exposure. Thus, PET has been extensively used for the diagnosis of various diseases as well as in drug discovery processes, particularly in human microdosing trials during the early stage of drug development (see also Section 5) [34, 35]. The potential applications of PET molecular imaging in an interdisciplinary scientific area strongly depend on the availability of suitable radioactive molecular probes with specific biological functions. The development of biologically significant and novel PET probes can be accomplished by combining an efficient synthetic strategy for designing molecules and new advances in the field of labeling chemistry [36]. Among the short-lived positron-emitting radionuclides, ^{11}C and ^{18}F, with a half-life of 20.4 and 109.8 min, respectively, have often been used for radiolabeling because they are the most significant radionuclides from both a chemical and biological perspective. These ^{11}C and ^{18}F-labeled precursors can be synthesized via nuclear reaction as shown in **Figure 8**.

Need for rapid reactions for ^{11}C-labeling: Certain aspects of PET radiochemistry such as short half-lives, extremely small amounts of available radionuclides, and relatively high-energy radiation, impose severe restrictions on the synthesis of PET probes. In particular, the synthesis of a pure, injectable ^{11}C-labeled probe must be accomplished within a maximum of two to three half-lives (i.e., 40–60min) because of its rapidly decaying radioactivity. The synthetic processes include: (1) preparation of a standard radiolabeled precursor such as $[^{11}C]CH_4$, $[^{11}C]CH_3I$, $[^{11}C]CH_3OTf$, $[^{11}C]CO$, and $[^{11}C]CO_2$ based on the nuclear reaction $^{14}N(p,\alpha)^{11}C$; (2) ^{11}C-labeling the synthesis of the target PET probe (see also **Figure 8**); (3) work-up and

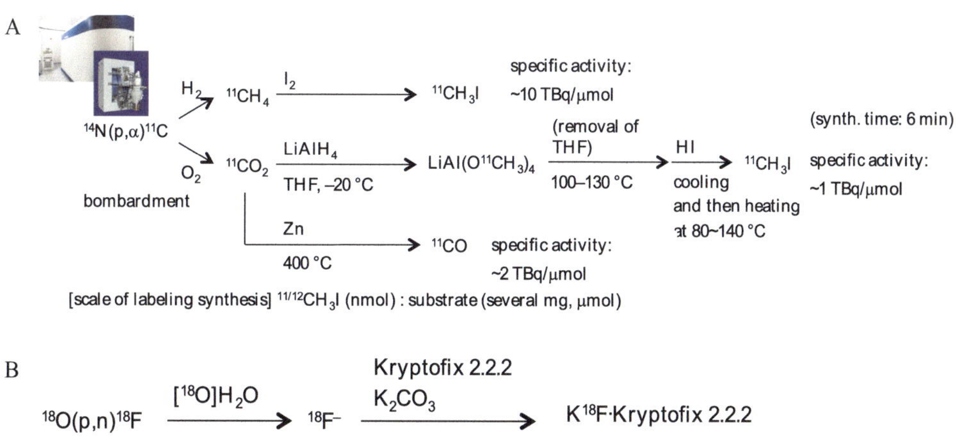

Figure 8. Production of ^{11}C and ^{18}F precursors from nuclear reaction. A. Preparation of $[^{11}C]$ carbon dioxide, $[^{11}C]$ carbon monoxide, and $[^{11}C]$ methyl iodide as ^{11}C-precursors; B. Preparation of $[^{18}F]$ fluoride ion.

chromatographic purification; and (4) preparation of an injectable solution for an animal/human PET study (pharmaceutical formulation). Thus, the time allowed for a ^{11}C-labeling reaction should be less than 5 min, inevitably necessitating a rapid chemical reaction. Another limitation encountered in the synthesis of a ^{11}C-labeled PET probe is the extremely small amounts (normally nanomol level) of the ^{11}C-labeling precursors such as [^{11}C]CH_3I. Therefore, the labeling reaction is usually carried out with a larger amount (several milligrams) of the reacting substrate (micromol level) to conduct the reaction conveniently. A PET probe thus obtained must be highly pure; the PET probe for clinical use should be synthesized at the Good Manufacturing Practice (GMP) level according to the regulation guidelines (also see Section 5).

Toward general ^{11}C-labelings, four types of rapid C-[^{11}C]methylations: In PET radiochemistry, the [^{11}C]methylation of hetero-atoms such as N, O, and S has mainly been explored because of its simple reaction conditions wherein [^{11}C]CH_3I is mixed with a large amount of the substrate. However, a carbon-hetero-atom bond tends to be metabolized to produce [^{11}C]CH_3OH, [^{11}C]CH_2O, and H[^{11}C]COOH, which are dispersed over the entire organ, thus decreasing the credibility of the PET image. In this study, we considered employing [^{11}C]methylation via the C—^{11}C bond formation (referred to as *C*-[^{11}C]methylation) [3, 4] as it has a number of advantages: (1) the [^{11}C]methyl group attached to the C atom is much more stable metabolically, thus providing a highly credible PET image; (2) a methyl group is the smallest non-polar substituent with the least influence on the parent biological activity; (3) short half-lived ^{11}C is favorable to rapidly screen optimized reaction conditions and ready evaluation of *in vivo* behavior allows several trials per day. These attractive features and benefits prompted us to investigate the realization of four types of rapid *C*-[^{11}C]methylations on organic frameworks involving arene, alkene, alkyne, and alkane, namely cross-coupling reactions between sp^2 (aryl) and sp^3 (alkyl) hybridized carbons, sp^2 (alkenyl) and sp^3 (alkyl) carbons, sp (alkynyl) and sp^3 (alkyl) carbons, and sp^3 (alkyl) and sp^3 (alkyl) carbons (**Figure 9**). This allowed us to validate the ^{11}C-labeling of the entire range of organic compounds. Subsequent pharmacokinetic (PK)/pharmacodynamics (PD) studies in *in vivo* systems using PET could offer a new methodology to

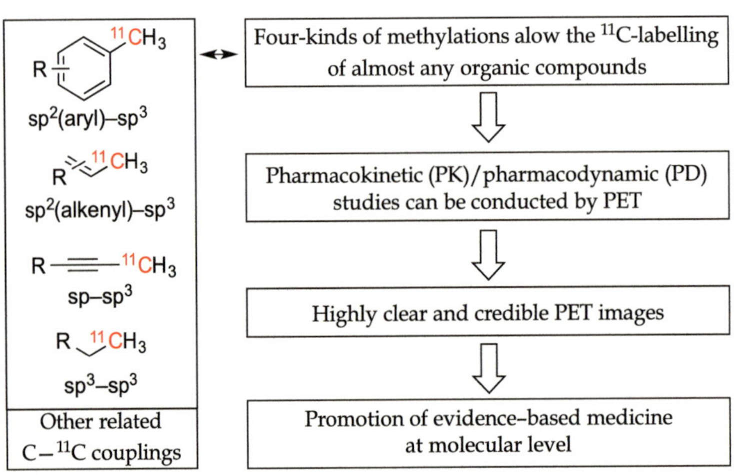

Figure 9. Four types of rapid *C*-[^{11}C]methylations and their benefits.

promote "evidence-based medicines" at the molecular level. The use of moisture-sensitive organolithium compound for this purpose could not justify the stoichiometry of an extremely small amount of [^{11}C]CH_3I, resulting in the unavoidable production of a large amount of an undesired demethylated derivative due to the use of an excess amount of the lithiated substrate together with the formation of undesired side reactions under such drastic conditions [37]. Accordingly, the tedious work to remove the demethylated side products is obliged to be done for obtaining the desired ^{11}C-labeled compound with high purity.

Benefits and problems of organostannanes as trapping substrates: The Stille reaction was considered as the *first choice* for this purpose (synthesis of PET probes using [^{11}C]CH_3I) because of the favorable properties of organostannanes [38] (organotin compounds): (1) easy preparation via a number of synthetic routes even when they contain a variety of reactive functional groups; (2) organostannanes are not particularly oxygen or moisture sensitive; (3) high tolerance to various chemical reactions and chromatographic purification conditions, enabling the incorporation of a radioisotope in the final or close to the final step; (4) extremely low polarity that allows an easy separation of the desired product from the substantial remaining tin substrate after labeling reaction. However, the Stille reaction was rarely used with methyl iodide as an sp^3-hybridized carbon partner, compared to its wide applicability to aryl or allyl halides with sp^2- or sp^3-hybridized carbons; it is rather difficult to realize methylation in high yields. Actually, the Stille group reported that the reaction of CH_3I and *p*-methoxyphenyltributylstannane in the presence of [Pd{P(C_6H_5)$_3$}$_4$] at 50°C for 24 h induced a scrambling reaction between the methyl and phenyl groups in the methyl iodide and triphenylphosphine, respectively, affording the desired *p*-methoxytoluene at only 3% yield along with undesired 1-methoxy-4-phenylbenzene as a byproduct at 8% yield [39]. The use of more reactive phenyltrimethylstannane as the substrate also induces competition between [^{11}C]CH_3 in [^{11}C]CH_3I and CH_3 and phenyl groups in the stannane to produce [^{11}C]CH_3CH_3 and non-radioactive $C_6H_5CH_3$ as byproducts (also see Section 3.1.1) [40]. In this study, we have considered devising new reaction conditions that are capable of promoting a rapid cross-coupling reaction.

3.1. Exploration of four kinds of rapid *C*-[^{11}C]methylations

3.1.1. *Optimized conditions to promote the rapid reaction of methyl iodide with an excess amount of phenyltributylstannane (rapid coupling between sp^2(aryl)-sp^3 hybridized carbons)*

Keeping in mind the ^{11}C radiolabeling conditions for a PET probe synthesis, we set up a model reaction using methyl iodide and an excess amount of phenyltributylstannane (**45**) ([CH_3I]/[**45**] = 1:40 in molar ratio) to possibly reduce the reaction time to less than 5 min (**Table 2**) [40]. The yield of the methylated product, toluene (**46**), was determined based on the CH_3I consumption. As expected, the conventional Stille reaction conditions with a reaction time of 30 min did not afford the desired product (**Table 2**, Entry 1), leading us to introduce the concept of coordinative unsaturation to activate the Pd catalyst. Thus, we found that the use of a coordinatively unsaturated Pd^0 complex, [Pd{P(*o*-$CH_3C_6H_4$)$_3$}$_2$] [41], generated *in situ* by mixing [Pd_2(dba)$_3$] and sterically bulky tri-*o*-tolylphosphine (cone angle = 194° for P(*o*-$CH_3C_6H_4$)$_3$) [42] instead of triphenylphosphine (cone angle = 145° for P(C_6H_5)$_3$) [42], significantly increased the yield (76%, Entry 2). Next, we introduced an additional concept to

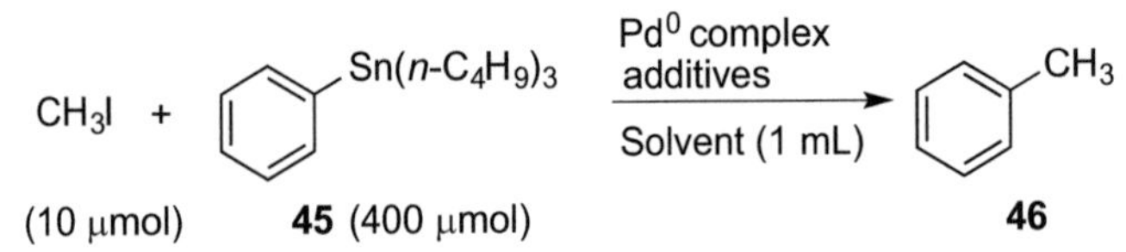

Entry	Pd^0 complex (μmol)	Ligand (L) and/or additive (μmol)	Solvent	Temp., °C	Time, min	Yield, %
1	$[Pd\{P(C_6H_5)_3\}_4]$ (10)	–	DMSO	40	30	0
2	$[Pd_2(dba)_3]$ (5)	$P(o\text{-}CH_3C_6H_4)_3$ (20)	DME	40	30	76
3	$[Pd_2(dba)_3]$ (5)	$P(o\text{-}CH_3C_6H_4)_3$ (20)	DME	80	5	41
4	$[Pd_2(dba)_3]$ (5)	$P(o\text{-}CH_3C_6H_4)_3$ (20)	DMF	80	5	63
5	$[Pd_2(dba)_3]$ (5)	$P(o\text{-}CH_3C_6H_4)_3$ (20), CuI (20)	DMF	60	5	3
6	$[Pd_2(dba)_3]$ (5)	$P(o\text{-}CH_3C_6H_4)_3$ (20), CuCl (20), K_2CO_3 (20)	DMF	60	5	91

Table 2. Rapid cross-coupling of methyl iodide and phenyltributylstannane (**45**).

shorten the reaction time (from 30 min to 5 min). Because simple heating (80°C) was less effective and lowered the yield, we stabilized the transiently formed Pd catalyst using DMF; this effectively increased the yield to a considerable extent (Entry 4). Furthermore, we tried to enhance the reaction rate by adding a Cu^I salt, expecting Sn to Cu transmetallation, and K_2CO_3 to react with the $(n\text{-}C_4H_9)_3SnX$ (X = I and/or Cl) byproduct generated during the reaction to neutralize the reaction mixture. Thus, the reaction of the CH_3I/**45**/$[Pd_2(dba)_3]$/$P(o\text{-}CH_3C_6H_4)_3$/CuCl/K_2CO_3 system (1:40:0.5:2:2:2) in DMF at 60°C for 5 min affords the desired product at 91% yield (**Table 2**, Entry 5) [40]. It should be noted that the reaction using phenyl*trimethyl*stannane affords (**46**) at 122–129% yield because of ethane impurity, indicating the undesired cross-coupling reactions (scrambling) between the CH_3 groups in CH_3I and those on the tin atom (phenyltrimethylstannane). Such a reaction between $[^{11}C]CH_3I$ and phenyltrimethylstannane under PET radiolabeling conditions produce undesired radioactive and volatile ethane, $[^{11}C]CH_3CH_3$. The product from the scrambling reaction between the phenyl and methyl groups on the tin atom also contaminate toluene (the undesired product in the actual PET probe synthesis) to a significant extent [40], decreasing the yield of the desired $[^{11}C]$(**46**); thereby reducing the molar activity [43]. These phenomena were also reported during the Pd-catalyzed reaction of 1-(2′-deoxy-2′-fluoro-β-D-arabinofuranosyl)-5-(trimethylstannyl)uracil to synthesize $[^{11}C]$FMAU [44]. Therefore, we concluded that phenyltributylstannane, although less reactive, would serve as a much more suitable coupling partner than phenyltrimethylstannane because of the increased yield of the reaction, radiation safety and relatively low toxicity: it should be noted that tributyltin derivative is practically non-toxic, while trimethyltin has a significant acute toxicity [45].

The reaction conditions thus discovered are significantly different from those of the originally reported Stille coupling. The rapid coupling of CH_3I and (**45**) probably proceeds by the mechanism proposed in Eqs. 1–5 [40].

$$\mathrm{CH_3I + [Pd\{P(\textit{o}\text{-}CH_3C_6H_4)_3\}_2] \longrightarrow \underset{\mathbf{47}}{[Pd(CH_3)I\{P(\textit{o}\text{-}CH_3C_6H_4)_3\}]} + P(\textit{o}\text{-}CH_3C_6H_4)_3} \quad (1)$$

$$\mathrm{\underset{\mathbf{45}}{C_6H_5Sn(\textit{n}\text{-}C_4H_9)_3} + CuX + P(\textit{o}\text{-}CH_3C_6H_4)_3 \rightleftharpoons \underset{\mathbf{48}}{[Cu(C_6H_5)\{P(\textit{o}\text{-}CH_3C_6H_4)_3\}]} + (\textit{n}\text{-}C_4H_9)_3SnX} \quad (2)$$

$$\mathrm{2\,(\textit{n}\text{-}C_4H_9)_3SnX + K_2CO_3 \longrightarrow \underset{\mathbf{49}}{[(\textit{n}\text{-}C_4H_9)_3SnO]_2C{=}O} + 2\,KX} \quad (3)$$

$$\mathrm{\mathbf{47} + C_6H_5M \longrightarrow \underset{\mathbf{50}}{[Pd(CH_3)(C_6H_5)\{P(\textit{o}\text{-}CH_3C_6H_4)_3\}]} + MI} \quad (4)$$

$$\mathrm{\mathbf{50} \longrightarrow \underset{toluene\ (\mathbf{46})}{H_3C{-}C_6H_5} + [Pd\{P(\textit{o}\text{-}CH_3C_6H_4)_3\}]} \quad (5)$$

$$\mathrm{X = Cl\ or\ Br;\ M = Sn(\textit{n}\text{-}C_4H_9)_3\ or\ Cu\{P(\textit{o}\text{-}CH_3C_6H_4)_3\}_{\textit{n}}.}$$

Thus, CH_3I undergoes oxidative addition with a Pd^0 species to afford methyl-Pd^{II} iodide (**47**) (oxidative addition, Eq. (1)). The Pd^{II} complex (**47**) may directly react with (**45**) to afford (methyl)(phenyl)Pd^{II} complex (**50**) (substitution, Eq. (4)); however, the formation of (**50**) would be facilitated via Cu complex (**48**) formed by the Sn/Cu transmetallation (Eq. (2)). K_2CO_3 reacts with (n-$C_4H_9)_3SnX$ to form a stable, neutral bis(tributylstannyl)carbonate (**49**) (Eq. (3)). At the same time, K_2CO_3 synergically works with a Cu^I salt to promote the Sn/Cu transmetallation (Eqs. (2) and (3)) [40]. Finally, (**46**) is formed by the reductive elimination of Pd^{II} complex (**50**) (Eq. (5)). The significant effect of tri-*o*-tolylphosphine is attributed to its considerable bulkiness, which facilitates the generation of the coordinatively unsaturated Pd^0 and Pd^{II} intermediates. The substitution reaction leading to (**50**) and/or the reductive elimination leading to (**46**) requires the formation of the coordinatedly unsaturated tricoordinate Pd^{II} complex. DMF may stabilize such Pd intermediates even at high temperatures. It was very difficult to conduct the Stille reaction using CH_3I as an sp^3-carbon partner [39] until our successful attempt in this field [3, 4]. Prabhakaran et al. used our reaction conditions for the synthesis of [^{11}C]celecoxib using a tributyltin substrate after protecting the sulfonamide group [46]. In this context, we promoted the ^{11}C-labeling without such a protection using the organoborane substrate, yielding higher efficiency for the synthesis (see Section 3.1.5.1.1) [47, 48].

3.1.2. Synthesis of 15R-[^{11}C]TIC methyl ester, a PET probe specific for prostacyclin receptor subtype (IP_2) in the central nervous system

During a study of the design and synthesis of a specific prostaglandin (PG) probe, we succeeded in developing (15*R*)-16-*m*-tolyl-17,18,19,20-tetranorisocarbacyclin (15*R*–TIC, (15*R*)-**27b**) with specifically high affinity to a novel prostacyclin receptor (IP_2) in the central nervous system (see also Section 2.4.1.2) [25, 49]. The tolyl group in (15*R*)-**27b** was considered as a trigger component to create the PET probe as illustrated in the retrosynthesis of 15*R*-[^{11}C]TIC (**Scheme 12**). The tin precursor ((15*R*)-**53b**) was synthesized as shown in **Scheme 13** [50]. The Horner-Emmons reaction of the aldehyde (**18**) and the phosphonate (**51**) gave the enone (**52**) at

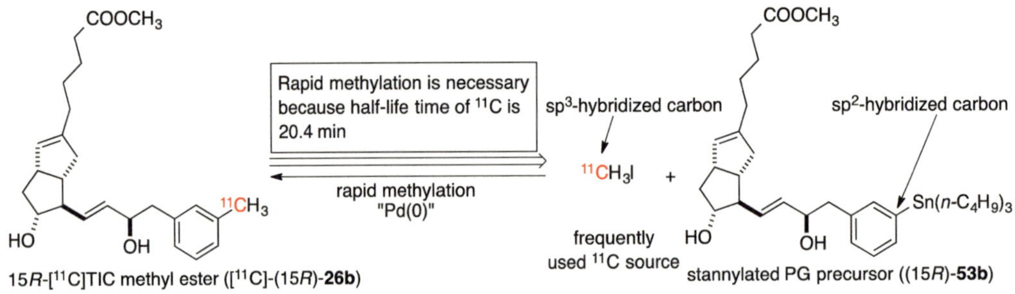

Scheme 12. Plan for ^{11}C-incorporation: retrosynthesis.

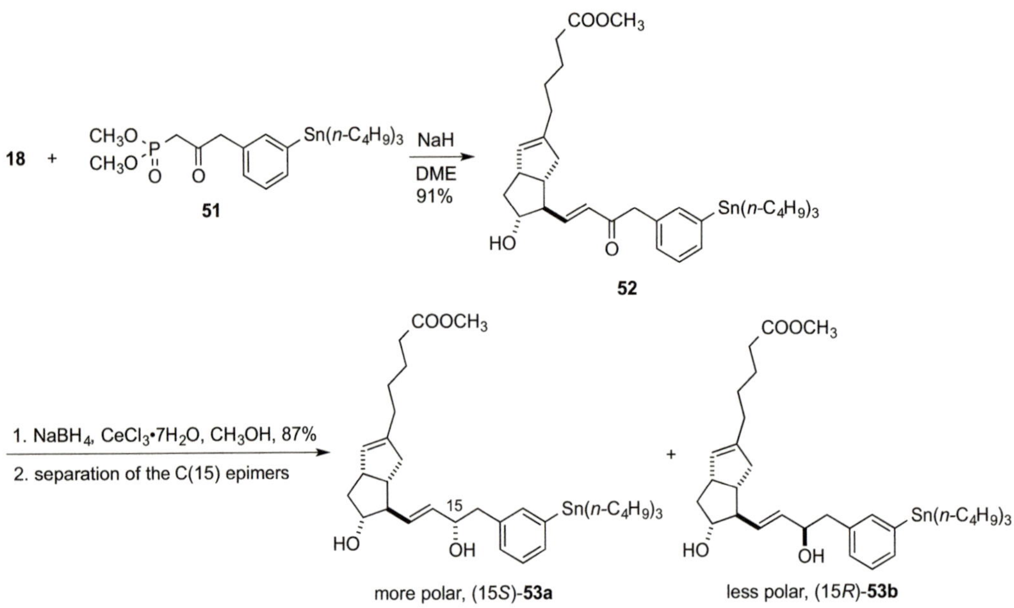

Scheme 13. Synthesis of (15*R*)-16-[3-(tri-*n*-butylstannyl)phenyl]-17,18,19,20-tetranorisocarbacyclin methyl ester ((15*R*)-**53b**).

91% yield. The chemoselective 1,2-reduction of the C(15) keto group with $NaBH_4$ in the presence of $CeCl_3$ gave a 1:1 mixture of the stereoisomeric diols at 87% yield. The C(15) stereoisomers were separated by conventional silica gel column chromatography, giving (15*S*)-**53a** and (15*R*)-**53b** as more and less polar materials, respectively. Thus, we intended to apply rapid *C*-methylation conditions to synthesize the 15*R*-[^{11}C]TIC methyl ester ([^{11}C]-(15*R*)-**26b**) using [^{11}C]CH_3I, prepared from [^{11}C]CO_2 according to an established method [51], and stannane (15*R*)-**53b** [50]. However, we found that *C*-[^{11}C]methylation under radiolabeling conditions lacked reproducibility for some unknown reasons even after using excess Cu^I salt. To overcome this difficulty, we found valuable information for the problem caused by CuI, which severely reduced the methylation rate of (**45**) (**Table 2**, Entry 5). In order to minimize this inhibitory effect of CuI, we changed the one-pot operation to a *two-pot stepwise procedure*

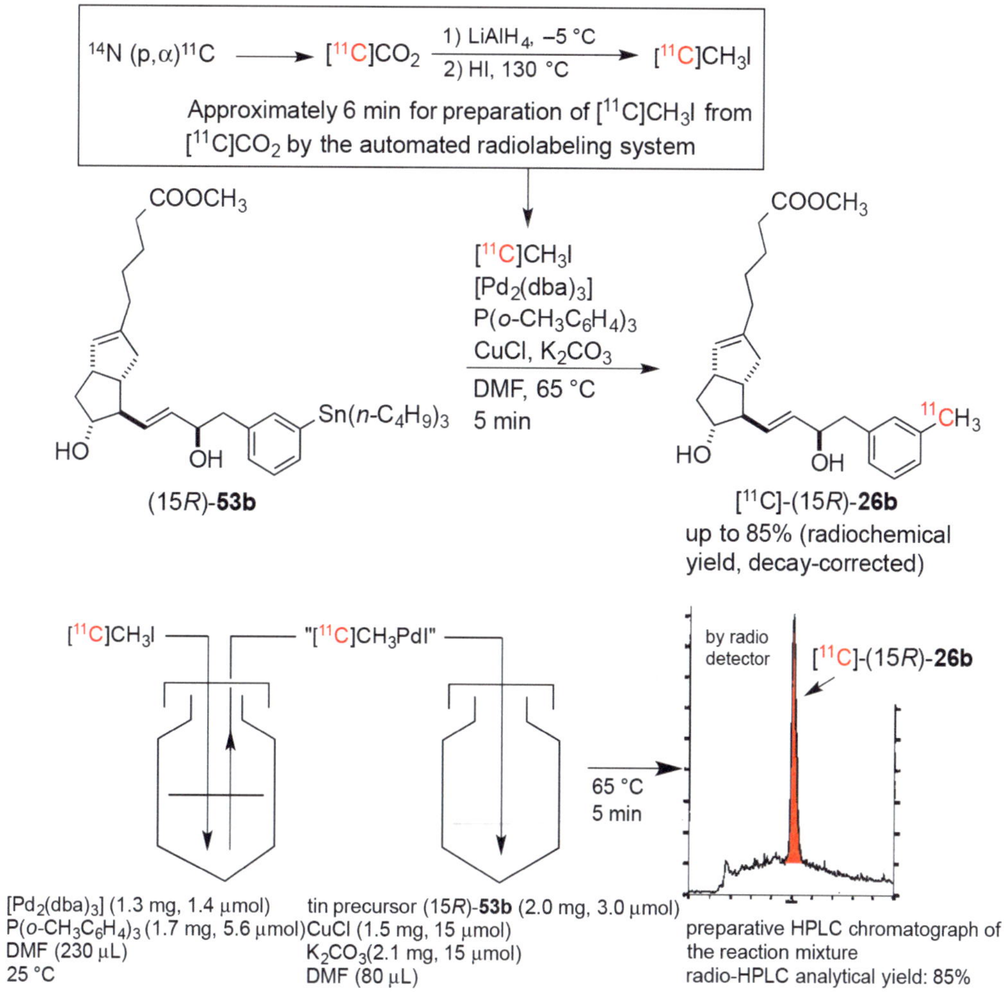

Figure 10. Synthesis of 15*R*-[^{11}C]TIC methyl ester ([^{11}C]-(15*R*)-**26b**) using a two-pot stepwise procedure.

for the actual PET probe synthesis (**Figure 10**) [2]. This procedure consists of independent syntheses of methylpalladium and a phenyl copper complex at room temperature (RT, 25°C), then a mixing of these species, and heating at an elevated temperature (65°C, 5 min). Thus, a highly sensitive PET probe, [^{11}C]-(15*R*)-**26b**, was obtained at up to 85% radiochemical yield (decay-corrected; total radioactivity of 2–3 GBq; greater than 98% radiochemical and chemical purities; total synthesis time, 35–40 min (radiochemical purity is a proportion of a magnitude of desired radiolabeled peak for all the radiolabeled peaks of the isolated product by analytical radio-HPLC; chemical purity is the proportion of desired peak for all the peaks of the isolated product by analytical UV-HPLC)), which was then used in rat, monkey, and then human PET studies.

Here, the principal author, M. Suzuki, became the first volunteer, after the approval of the Ethics Committee. Thus, [^{11}C]-(15*R*)-**26b** was injected into his right arm, which then passed

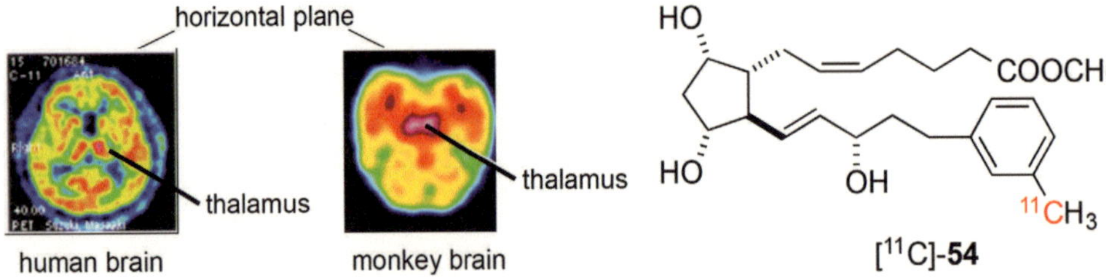

Figure 11. PET imaging in human (left) and monkey (right) brains using (15R)-[^{11}C]TIC-Me ([^{11}C]-(15R)-**26b**).

through the blood–brain barrier to be hydrolyzed in the brain. The resulting free carboxylic acid eventually bound to the IP_2 receptor distributed throughout various parts of the human brain (**Figure 11**, see also **Figure 7**) [2]. A PET study of the middle cerebral artery occlusion using a monkey model demonstrated that 15*R*–TIC has a potent neuroprotective effect against focal cerebral ischemia as judged by the [^{15}O]O_2 consumption and the uptake of [^{18}F]FDG [52]. The PET studies for rats and humans using the [^{11}C]-(15*R*)-**26b** showed that it could be an useful probe for the *in vivo* analyses of the MRP2-mediated hepatobiliary transport [53], 17-(3-[^{11}C]Methylphenyl)-18,19,20-trinor-prostaglandin $F_2\alpha$ isopropyl ester ([^{11}C]-**54**) [54] targeting the $PGF_2\ \alpha$ receptor was also synthesized under similar reaction conditions.

3.1.3. Rapid C-methylation of heteroaryl-substituted stannanes and application to ^{11}C-labeling

There is a great demand to incorporate short-lived ^{11}C-labeled methyl groups into heteroaromatic carbon structures because such structures often appear in major drugs and their promising drug candidates. The rapid methylations of hetero-aromatic ring-substituted tributylstannanes (**55a–i**) were carried out [55] by first employing our previous reaction conditions: CH_3I/**55a–i**/[$Pd_2(dba)_3$]/P(*o*-$CH_3C_6H_4)_3$/CuCl/K_2CO_3 (1:40:0.5:2:2:2) in DMF at 60°C for 5 min (conditions A) [55], affording the desired products in low yields (**Table 3**). An increase in the phosphine ligand (conditions B) significantly improved the yields of (**56b**), (**56c**), and (**56i**); however, the conditions were still insufficient in terms of adaptable heteroaromatic structures (substrate scope). Another CuBr/CsF combination system (conditions C) also provided the same results as conditions B. Pyridine and the related heteroaromatic compounds still remained less reactive. Consequently, the problem was solved by replacing DMF with NMP. Interestingly, the solvent effect though not observed for the CuCl/K_2CO_3 system, was observed for the CuBr/CsF system (**Table 4**, Entry 2), affording 2-methylpyridine (**56d**) at 81% yield. Other solvents except amide-type solvents and amine additives were not effective. Thus, the reaction in NMP at 60–100°C for 5 min using CH_3I/**55a–i**/[$Pd_2(dba)_3$]/P(*o*-$CH_3C_6H_4)_3$/CuBr/CsF (1:40:0.5:16:2:5) (conditions D) afforded the methylated products (**56a–i**) at >80% yields for all the heteroaromatic compounds listed in **Table 3**. It should be noted that we did not find the conditions using a [Pd{P(*tert*-$C_4H_9)_3\}_2$]/CsF system in NMP reported by Fu [56] effective for methylation (21% yield, **Table 4**, Entry 2).

2- and 3-[^{11}C]Methylpyridines ([^{11}C]-**56d** and **56e**) were obtained at 88% and 91% radio-HPLC analytical yields (a percentage of the magnitude of the peak of the desired product for all the

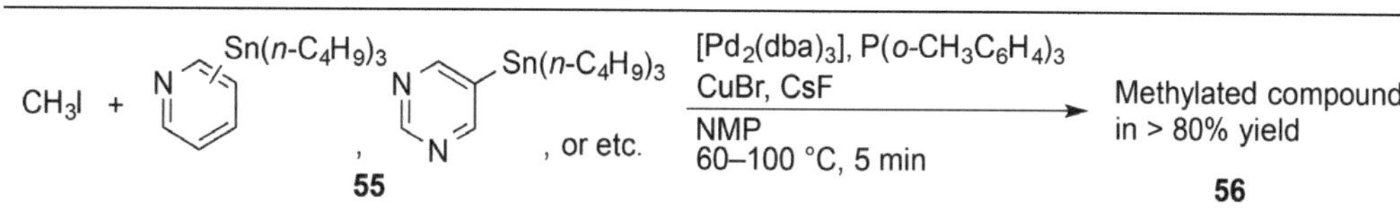

Entry	Heteroaryl stannane 55	Methylated compound 56	Yield, %			
			A[a]	B[a]	C[a]	D[a]
1	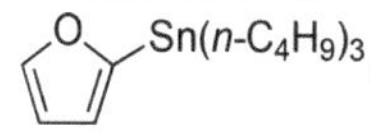 55a	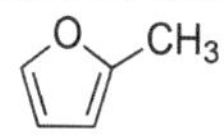 56a	28	75	73	**80**
2	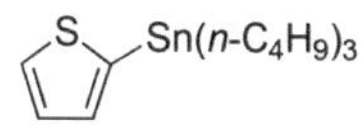 55b	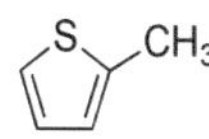 56b	57	**87**	**91**	**94**
3	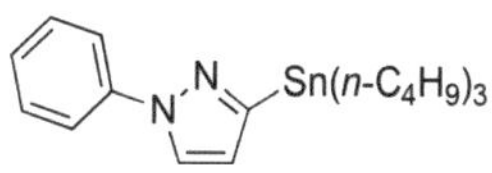 55c	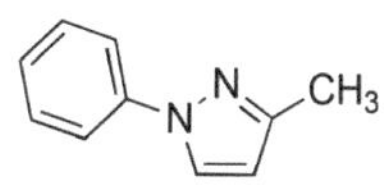 56c	52	**88**	**90**	**94**
4	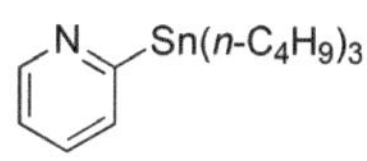 55d	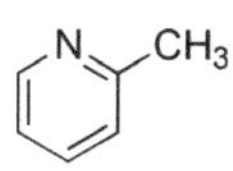 56d	16	67	63	**81**
5	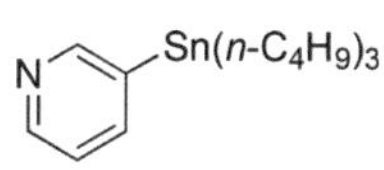 55e	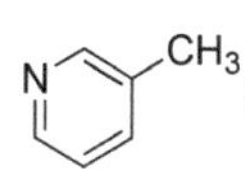 56e	25	61	66	**80**
6	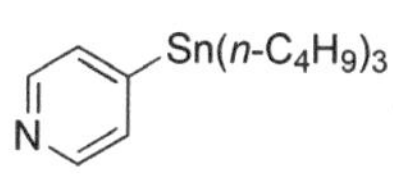 55f	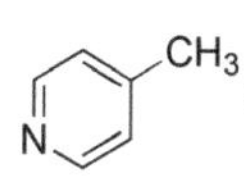 56f	79	60	68	**87**
7	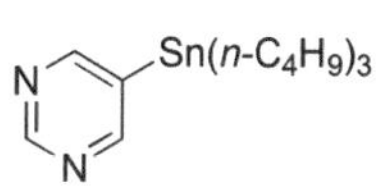 55 g	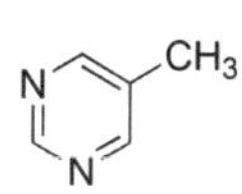 56 g	3	50	48	**87**[b]
8	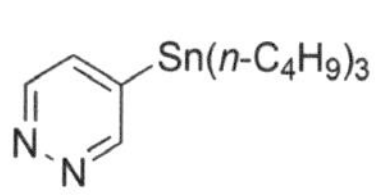 55 h	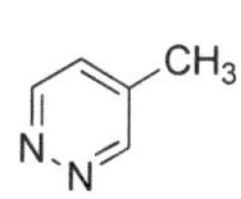 56 h	25	72	70	**90**
9	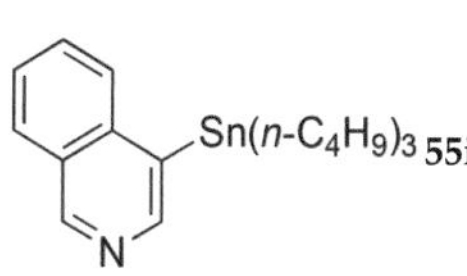 55i	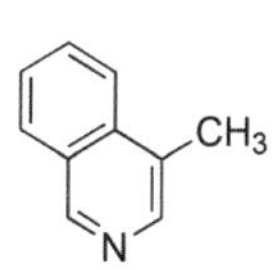 56i	12	**83**	75	**83**

[a]Reaction conditions (molar ratio): A: CH_3I/ **55**/$[Pd_2(dba)_3]$/$P(o\text{-}CH_3C_6H_4)_3$/CuCl/K_2CO_3 (1:40:0.5:2:2:2) in DMF at 60°C for 5 min; B: CH_3I/**55**/$[Pd_2(dba)_3]$/$P(o\text{-}CH_3C_6H_4)_3$/CuCl/K_2CO_3 (1:40:0.5:16:2:5) in DMF at 60°C for 5 min; C: CH_3I/**55**/$[Pd_2(dba)_3]$/$P(o\text{-}CH_3C_6H_4)_3$/CuBr/CsF (1:40:0.5:16:2:5) in DMF at 60°C for 5 min; D: CH_3I/**55**/$[Pd_2(dba)_3]$/$P(o\text{-}CH_3C_6H_4)_3$/CuBr/CsF (1:40:0.5:16:2:5) in NMP at 60°C for 5 min.
[b]The reaction was conducted at 100°C.

Table 3. General rapid *C*-methylation on various neutral and basic heteroaromatic rings.

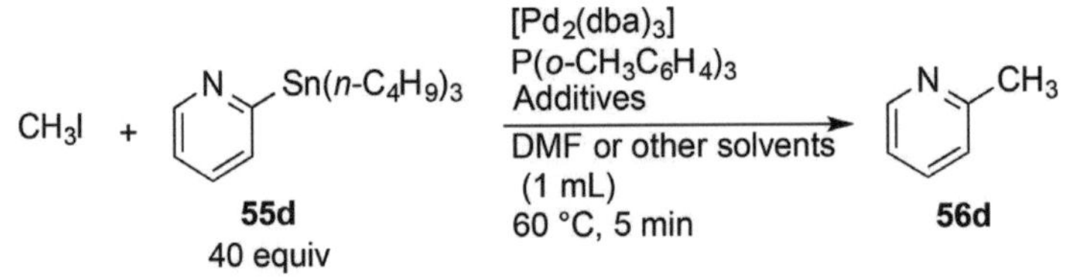

CH_3I/55d/[$Pd_2(dba)_3$]/P(o-$CH_3C_6H_4)_3$/CuCl/K_2CO_3 (1:40:0.5:16:2:2) or CH_3I/55d/[$Pd_2(dba)_3$]/P(o-$CH_3C_6H_4)_3$/CuBr/CsF (1:40:0.5:16:2:5) at 60°C for 5 min.

Entry	Solvent	Additive, equiv	Yield, %	
			CuCl/K_2CO_3	CuBr/CsF
1	DMF	—	67	65
2	NMP	—	66	**81** (21)[a] (39)[b]
3	DMA	—	—	69
4	DMI	—	—	18
5	toluene	—	—	20
6	THF	—	—	38
7	DMSO	—	—	23
8	HMPA	—	—	34
9	DMF	2,6-lutidine (17)	—	19
10	DMF	triethylamine (14)	—	20
11	DMF	DABCO(18)	—	6

[a]Fu's original conditions [56] (molar ratio): CH_3I/**55d**/[Pd{P(*tert*-$C_4H_9)_3\}_2$]/CsF (1:40:1:2).
[b]Fu's original conditions + CuBr (molar ratio): CH_3I/**55d**/[Pd{P(*tert*-$C_4H_9)_3\}_2$]/CuBr/CsF (1:40:1:2:5).

Table 4. Effect of solvent and additive with increased phosphine and synergic system (CuCl/K_2CO_3 or CuBr/CsF) on the rapid coupling of methyl iodide and 2-pyridyltributylstannane (**55d**) to afford 2-methylpyridine (**56d**).

peaks of radiolabeled compounds appearing in the radio-HPLC chart), respectively, using the [$Pd_2(dba)_3$]/P(o-$CH_3C_6H_4)_3$/CuBr/CsF (1:16:2:5) system in NMP at 60°C for 5 min (**Figure 12**) [55].

Synthesis of [^{11}C]H-1152, a PET probe specific for Rho-kinases: H-1152 (**60**) is known as the most potent, specific, and membrane-permeable inhibitor of small G protein Rho-associated kinase (Rho-kinase). A ^{11}C-labeled H-1152 ([^{11}C]-**60**), as a novel PET probe, for imaging Rho-kinases was synthesized for the first time based on the Pd^0-mediated rapid C-[^{11}C]methylation for heteroaryl-substituted stannane with a trifluoroacetyl (TFA) protecting group using [^{11}C]CH_3I followed by rapid deprotection of the TFA group [57]. The lithiation of bromide (**58**) prepared from (**57**) by protection with TFA, which is necessary for the subsequent tributylstannylation, was accomplished in a highly selective manner even in the presence of highly sensitive (to a nucleophile) TFA amide functional group. We focused on the use of *tert*-C_4H_9Li with stronger electron transfer capability and less nucleophilic character than *n*-C_4H_9Li and HMPA, which can modulate the structure of an organolithium compound by strong coordination. The

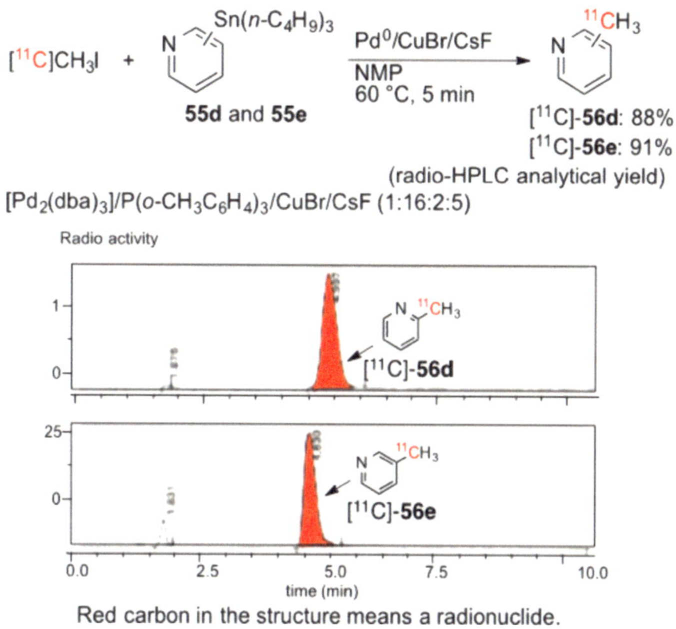

Figure 12. Syntheses of 2- and 3-[^{11}C]methylpyridines ([^{11}C]-**56d** and [^{11}C]-**56e**).

separated ion pair species formed under such HMPA-rich conditions [58], where the excess HMPA fully coordinate with the lithium cations and thereby reduce the coordination of Li to the carbonyl functional group, resulted in the loss of the reactivity of nucleophiles for the carbonyl to perform the stannylation highly selectively. Thus, the treatment of (**58**) with *tert*-C_4H_9Li (2 equiv) at −98°C for 2 h, followed by the continuous addition of HMPA (6 equiv) and (n-$C_4H_9)_3SnCl$ (1.5 equiv) at −98°C, then warming to RT for 10 h, afforded (**59**) at 68% yield (**Scheme 14**). *C*-[^{11}C]Methylation using [$Pd_2(dba)_3$]/P(o-$CH_3C_6H_4)_3$/CuBr/CsF (1:16:2:5) system in NMP at 80°C for 5 min followed by rapid deprotection using 2 M NaOH at 25–50°C for 1 min afforded [^{11}C]-**60** at 86% radio-HPLC analytical yield [57]. Isolated radioactivity: 3.8 GBq; radiochemical yield: 63% (decay-corrected); chemical and radiochemical purities: >99%; total synthesis time: 38 min. This PET probe can be used for the molecular imaging studies of cardiovascular diseases.

Syntheses of [methyl-^{11}C]thymidine, 4′-[methyl-^{11}C]thiothymidine, and [^{11}C]zidovudine ([^{11}C]AZT): In order to develop a more specific tumor imaging agent than [^{18}F]FDG [59, 60], the imaging studies were focused on the phosphorylation by thymidine kinase 1 (TK_1), whose activity increases by almost 10-fold during active DNA synthesis and proliferation of cancer [60]. 4′-Thiothymidine (**62b**), which shows similar biological properties as thymidine (**62a**) and a higher stability for the nucleoside cleavage than (**62a**), underwent ^{11}C-labeling and was used for tumor imaging in a rat where it exhibited a higher potential as an attractive PET probe than [^{18}F]FLT [61]. In this context, FMAU and (**62b**) have so far been labeled by ^{11}C using the 5-*trimethyl* and/or *tributyl*stannyl precursors [44]. However, such conditions resulted in lower efficiency for the reactions of 5-tributylstannyl-2′-deoxyuridine (**61a**) and 5-tributylstannyl-4′-thio-2′-deoxyuridine (**61b**) using [$Pd_2(dba)_3$]/P(o-$CH_3C_6H_4)_3$ (1:4) at 130°C for 5 min in DMF,

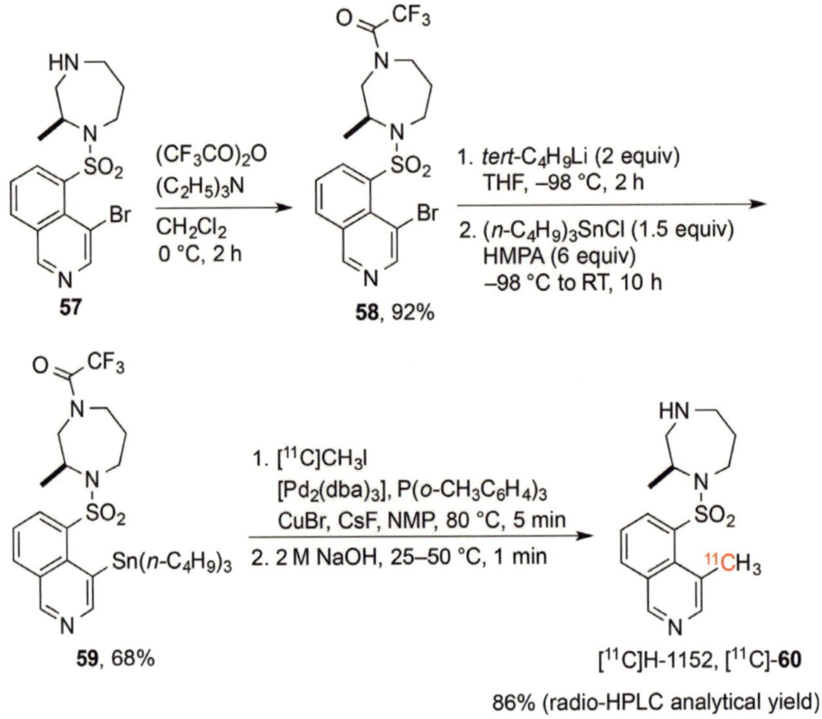

Scheme 14. Synthesis of [^{11}C]H-1152 ([^{11}C]-**60**), a PET probe specific for Rho-kinases.

affording the desired products (**62a**) and (**62b**) at 32% and 30% yields, respectively (**Table 5**, Entries 1 and 4). We tried to employ our reaction conditions here. The chemo-response of the thiothymidine precursor was different from the thymidine system [62]. The reaction using the CH_3I/**61a**/[$Pd_2(dba)_3$]/P(*o*-$CH_3C_6H_4)_3$/CuBr/CsF (1:25:1:32:2:5) system including another CuBr/CsF system, afforded (**62a**) at quantitative yield (Entry 3). The conditions for (**61b**) afforded (**62b**) at only 40% yield (**Table 5**, Entry 5). The reaction using five-fold amounts of CuBr/CsF at 80°C afforded (**62b**) at a much higher yield (83%, Entry 6); however, the reaction produced a large amount of an undesired destannylated product. Considering that the destannylated product (**64**) is produced by a proton transfer to the transmetallated Cu intermediate (**63**) from (**61b**$_{Cu2}$) with the enhanced acidity of the NH proton caused by Cu^{I} coordination to an S atom (**Figure 13**), we changed the medium to a much more basic system using a CH_3I/**61b**/[$Pd_2(dba)_3$]/P(*o*-$CH_3C_6H_4)_3$/CuCl/K_2CO_3 (1:25:1:32:2:5) system at 80°C to afford (**62b**) in nearly quantitative yield (98%, **Table 5**, Entry 7) [62]. Here, K_2CO_3 would prevent the proton transfer from (**61b**$_{Cu2}$) to (**63**) by forming (**65**).

Each optimized condition obtained for (**61a**) and (**61b**) was successfully used for the syntheses of the corresponding PET probes using the two-pot procedure (see Section 3.1.2) [62], affording [^{11}C]-**62a** and [^{11}C]-**62b** at 87 and 93% radio-HPLC analytical yields, respectively (**Figure 14**) [62]. The desired ^{11}C-labeled compounds were isolated by preparative HPLC to afford 45% in 42–59% isolated radiochemical yields (decay-corrected). The total synthesis time was 42 min in each case until radiopharmaceutical formulation with the total radioactivity as follows: 3.7–3.8 GBq; chemical purity: ≥98% and radiochemical purity: ≥99.5% sufficient for the human PET study [62].

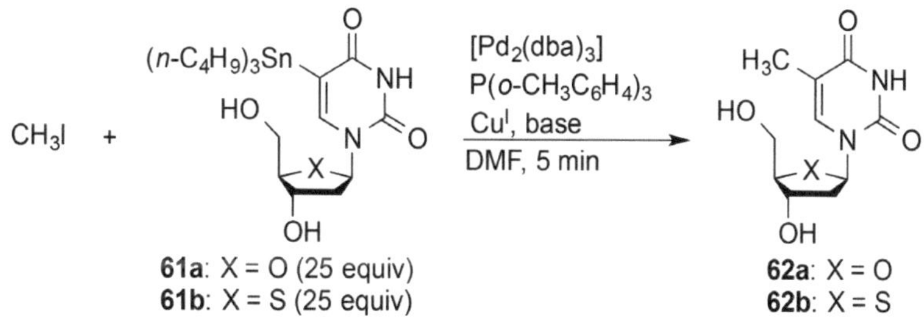

Entry	X	$P(o\text{-}CH_3C_6H_4)_3$, equiv	Cu^I, base (molar ratio)	Yield, %			
				60 °C	80 °C	100°C	130°C
1	O	4	–	0	–	–	32
2	O	32	CuCl, K_2CO_3 (2:5)	67	85	–	–
3	O	32	CuBr, CsF (2:5)	**100**	–	**97**	–
4	S	4	–	–	–	–	30
5	S	32	CuBr, CsF (2:5)	40	–	–	–
6	S	32	CuBr, CsF (10:25)	64	83	–	–
7	S	32	CuCl, K_2CO_3 (2:5)	83	**98**	–	–

Table 5. Synthesis of thymidine (**62a**) and 4′-thiothymidine (**62b**) by the rapid coupling of methyl iodide and 5-tributylstannyl-2′-deoxyuridine (**61a**) and 5-tributylstannyl-4′-thio-2′-deoxyuridine (**61b**).

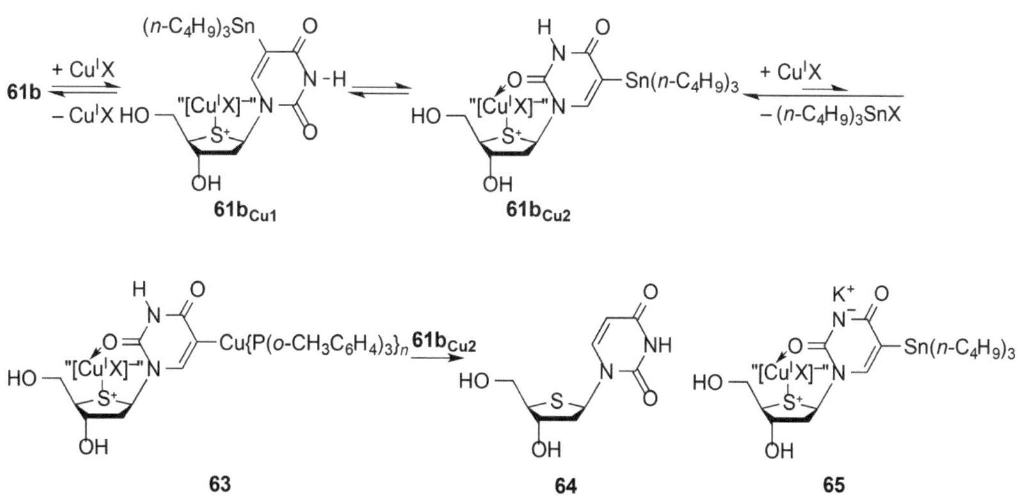

Figure 13. Assumed equilibration between a stannyl thiothymidine (**61b**) and a Cu^I salt.

One of reasons for the low reproducibility of the one-pot reaction for Pd^0-mediated rapid C-[^{11}C]methylation (see Section 3.1.2) using a stannyl precursor is considered to be that the highly reactive organocopper R—Cu (R = phenyl, vinyl, and heteroaromatic) species formed

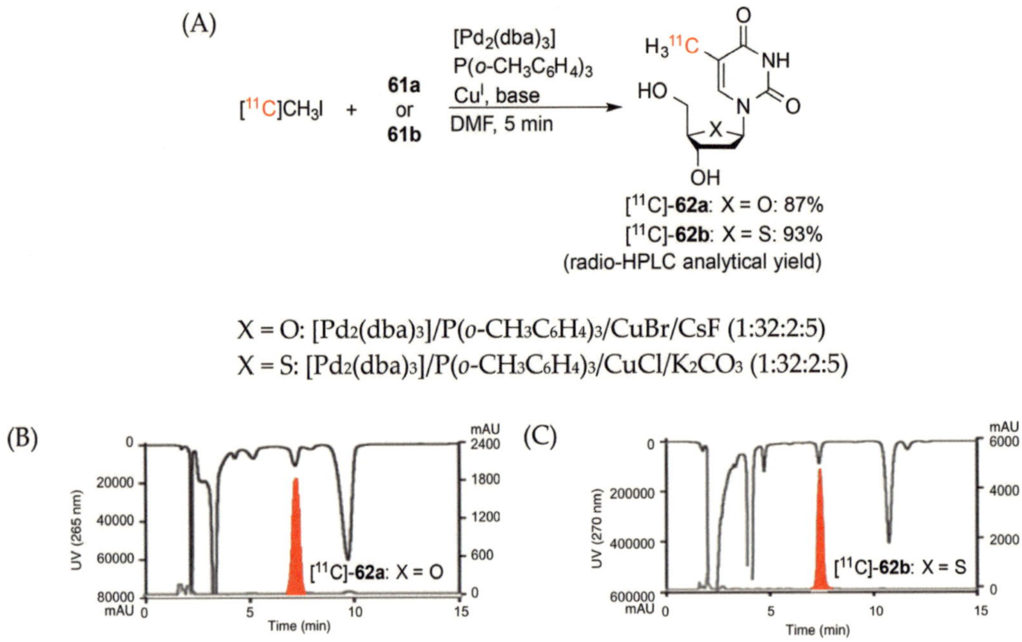

Figure 14. Synthetic scheme of [methyl–^{11}C]thymidine ([^{11}C]**62a**) or 4′-[methyl–^{11}C]thiothymidine ([^{11}C]**62b**) (A), and radio-HPLC analyses of [^{11}C]**62a** and [^{11}C]**62b** (B and C, respectively, radioactivity and UV vs. time). The peaks at the retention times of 7.6 and 7.4 min in B and C are [^{11}C]**62a** and [^{11}C]**62b**, respectively. (A) X = O: [$Pd_2(dba)_3$]/P(*o*-$CH_3C_6H_4)_3$/CuBr/CsF (1:32:2:5).

by the Sn/Cu transmetallation induces side reactions such as demetallation and homocoupling leading to RH and R—R, respectively. In this context, we found that the homocoupling of organocopper, R—Cu, can be greatly suppressed by bubbling [^{11}C]CH_3I in the mixture at lower temperature (−20°C) followed by heating at 80°C for 5 min. Thus, the efficient synthesis of [^{11}C]-**62b** was possible by the one-pot method, with much higher (double) isolated total radioactivity (7.6 GBq) [63]. The novel one-pot procedure was also applied to the ^{11}C-labeling of AZT (**67**), known as an anti-HIV and anti-tumor agent. Thus, the ^{11}C-labeling was conducted for (**66**) as shown in **Figure 15** to give [^{11}C]-**67**. The isolated [^{11}C]-**67** showed a high radioactivity of 7.0 GBq with a radiochemical purity of >99.5% [63]. In the PET experiment, [^{11}C]-**67** was shown to be accumulated in the C6 tumor with high selectivities (**Figure 15**) [64].

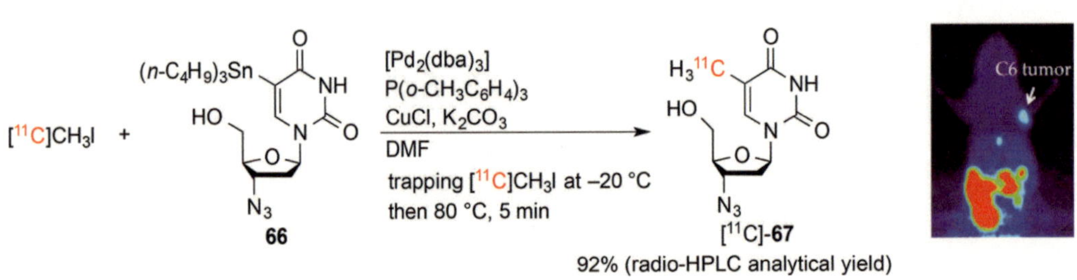

Figure 15. Synthetic scheme of [^{11}C]zidovudine ([^{11}C]AZT, [^{11}C]-**67**) and PET imaging of C6 tumor-bearing mice.

3.1.4. Rapid C-methylation of alkenes (rapid coupling between sp^2(vinyl)-sp^3 hybridized carbons) and application to ^{11}C-labeling

In the process of optimizing the rapid *C*-methylation conditions using 12 types of non-functional and functional 1-alkenyltributylstannanes (**68a–l**) (**Table 6**) as the model compounds, we developed generally applicable conditions, affording the corresponding methylated compounds (**69a–l**) at 90% or greater yields [65]. The high efficiency of the reaction: CH_3I/**68**/[$Pd_2(dba)_3$]/P(*o*-$CH_3C_6H_4)_3$/CuBr/CsF (1:40:0.5:2:2:5) is probably due to the use of a bulky arylphosphine, the synergic effect of the Cu^I salt and fluoride anions (Eqs. (2′) and (6)), which promotes the Sn/Cu transmetallation by shifting.

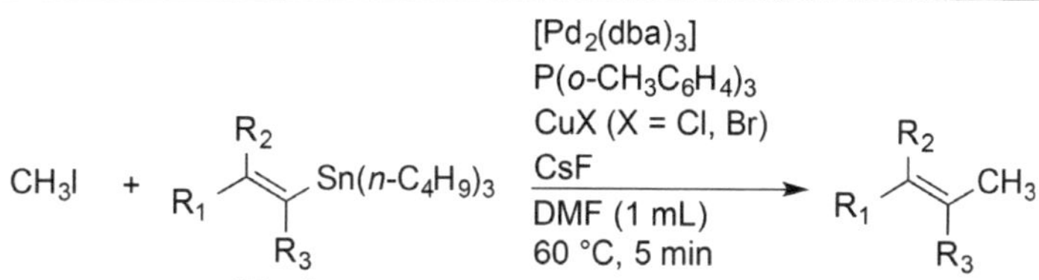

68a–l **69a–l**

CH_3I/68a–l/[$Pd_2(dba)_3$]/P(*o*-$CH_3C_6H_4)_3$/CuBr/CsF (1:40:0.5:2:2:5).

Entry	1-Alkenyltributylstannane 68	Yield, %	Entry	1-Alkenyltributylstannane 68	Yield, %
1	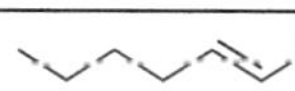 **68a**	98	7	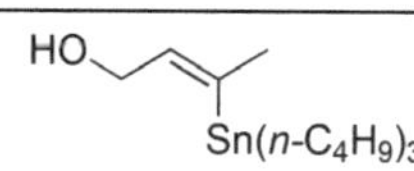 **68 g**	99
2	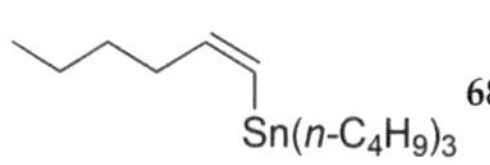 **68b**	99	8	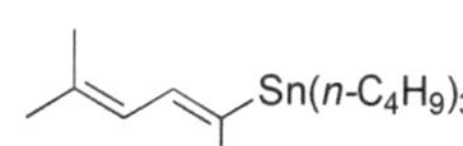 **68 h**	89 (91)[a]
3	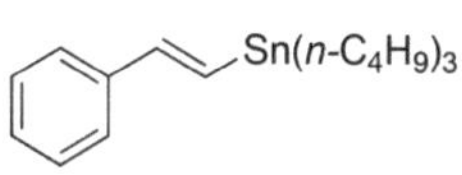 **68c**	83 (88)[a]	9	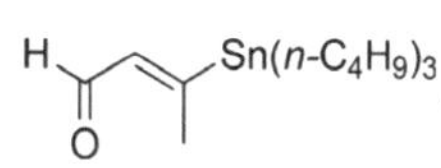 **68i**	85 (96)[a]
4	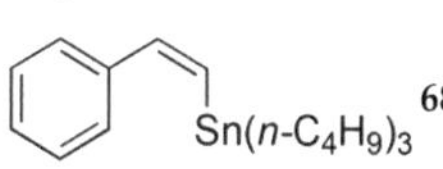 **68d**	84 (89)[a]	10	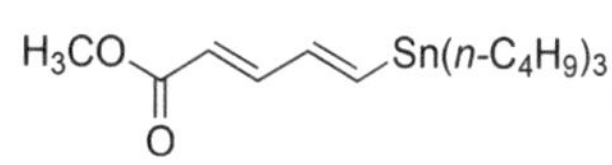 **68j**	86 (93)[a]
5	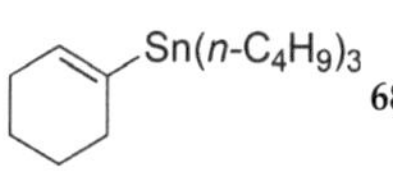 **68e**	99	11	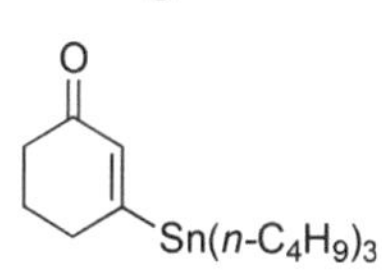 **68 k**	95
6	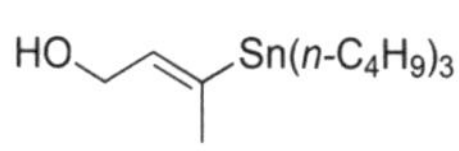 **68f**	99	12	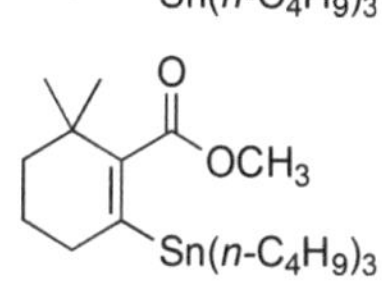 **68 l**	84 (91)[a]

[a]Modified conditions: [$Pd_2(dba)_3$]/P(*o*-$CH_3C_6H_4)_3$/CuBr/CsF (1:8:4:10).

Table 6. Rapid C-methylation on alkenyl structures.

$$\underset{\mathbf{68}}{RSn(n\text{-}C_4H_9)_3} + CuX + P(o\text{-}CH_3C_6H_4)_3 \rightleftharpoons [RCu\{P(o\text{-}CH_3C_6H_4)_3\}] + (n\text{-}C_4H_9)_3SnX \quad (2')$$

$$(n\text{-}C_4H_9)_3SnX + CsF \longrightarrow (n\text{-}C_4H_9)_3SnF\downarrow + CsX \qquad (6)$$

R = vinyl.

the equilibrium. The reactions using the Pd^0 complex, $[\{(\pi\text{-allyl})PdCl\}_2]/3P(tert\text{-}C_4H_9)_2(CH_3)$, $(CH_3)_4NF$, 3-Å molecular sieves [66]; or $[PdCl_2]/2P(tert\text{-}C_4H_9)_3$, CuI, CsF [67] afforded the desired products at only the same 2% yields using (**68e**). The stereoisomerization of the double bond did not occur at all.

The utility of the reaction was demonstrated by the synthesis of a ^{11}C-labeled partial retinoid derivative, [^{11}C]-**69l**, using reaction conditions with $CuCl/K_2CO_3$ or CuBr/CsF at 85% radio-HPLC analytical yields (**Figure 16**) [65].

3.1.5. Rapid C-methylation of alkynes (rapid coupling between sp-sp^3 hybridized carbons) and application to ^{11}C-labeling

As shown in **Table 7**, the reaction of tributyl-1-hexynyl-stannane (**70**) ($[CH_3I]/[\mathbf{70}] = 1{:}40$) using $[Pd\{P(C_6H_5)_3\}_4]$ for 5 min afforded the desired 2-heptyne (**71**) at a poor yield (Entry 1) [68]. Our previous conditions (described in Section 3.1.1), $[Pd_2(dba)_3]/P(o\text{-}CH_3C_6H_4)_3/CuCl/K_2CO_3$, were

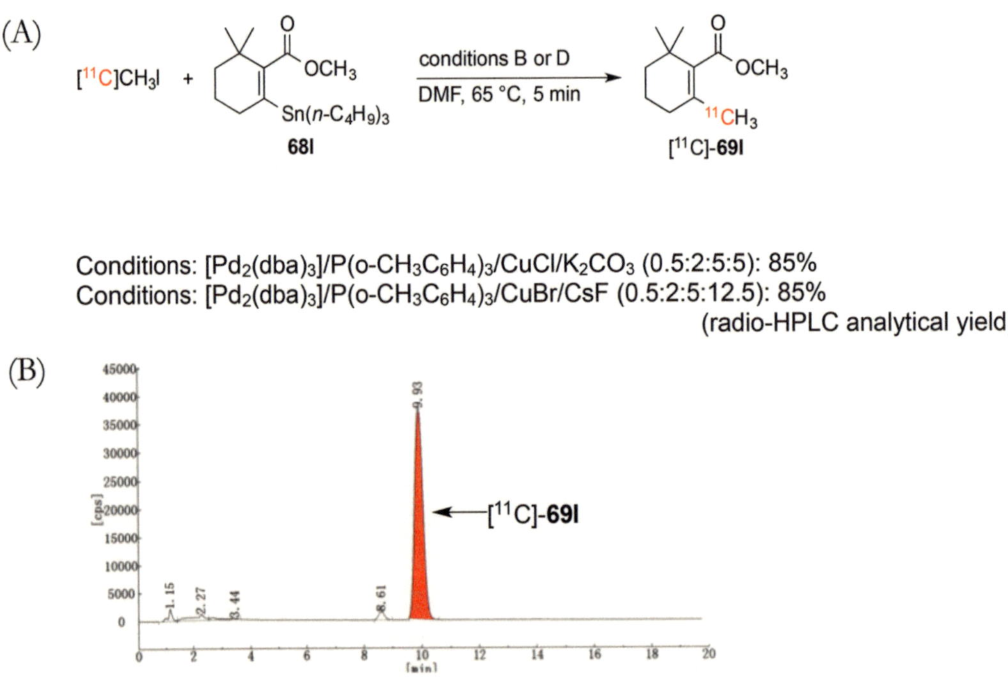

Figure 16. Synthesis of a ^{11}C-labeled partial stannane of retinoid [^{11}C]-**69 l** (A) and radio-HPLC analysis (B). Conditions for (A): $[Pd_2(dba)_3]/P(o\text{-}CH_3C_6H_4)_3/CuCl/K_2CO_3$ (0.5:2:5:5): 85%.

http://dx.doi.org/10.5772/intechopen.72868

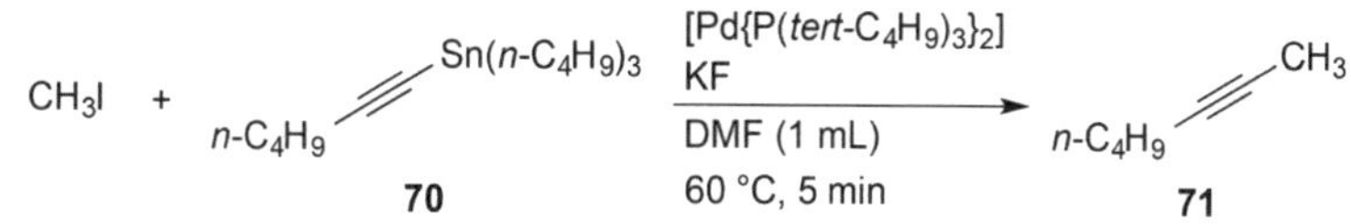

Entry	Pd^0	Additives	Yield, %
1	$[Pd(PPh_3)_4]$	—	27
2	$[Pd_2(dba)_3]$	PPh_3	11
3	$[Pd_2(dba)_3]$	P(2-furyl)$_3$	30
4	$[Pd_2(dba)_3]$	$AsPh_3$	61
5	$[Pd_2(dba)_3]$	$P(o\text{-}CH_3C_6H_4)_3$	20
6	$[Pd_2(dba)_3]$	$P(o\text{-}CH_3C_6H_4)_3$, CuCl, K_2CO_3	43
7	$[Pd_2(dba)_3]$	P(cyclohexyl)$_3$	49
8	$[Pd_2(dba)_3]$	$PCH_3(tert\text{-}C_4H_9)_2$	67
9	$[Pd_2(dba)_3]$	$P\ CH_3(tert\text{-}C_4H_9)_2$,CsF	10
10	$[Pd\{P(tert\text{-}C_4H_9)_3\}_2]$	$P(tert\text{-}C_4H_9)_3$	72
11	$[Pd\{P(tert\text{-}C_4H_9)_3\}_2]$	—	81
12	$[Pd\{P(tert\text{-}C_4H_9)_3\}_2]$	CuCl, K_2CO_3	45
13	$[Pd\{P(tert\text{-}C_4H_9)_3\}_2]$	CsF	92
14	$[Pd\{P(tert\text{-}C_4H_9)_3\}_2]$	KF	95
15	$[Pd\{P(tert\text{-}C_4H_9)_3\}_2]$	TBAF	15

Table 7. Pd^0-mediated rapid *C*-methylation using CH_3I and an excess amount of tributyl-1-hexynyl-stannane (**70**).

also not applicable (Entry 6). Consequently, we found that the bulky and strong σ electron-donating ligand, $P(tert\text{-}C_4H_9)_3$, facilitates the methylation [68]. The use of fluoride ions such as CsF or KF was extremely efficient in promoting the reaction. Thus, the reaction in the presence of $[Pd\{P(tert\text{-}C_4H_9)_3\}_2]$ and KF in DMF at 60°C for 5 min afforded **(71)** at 95% yield (Entry 14) [68]. The hypervalent stannate (**72**) is expected to be formed as a reactive species (Eq. (7)). The reaction was applicable to syntheses of 17β-estradiol and 5-(1-propynyl)-2′-deoxyuridine, which contains several types of functionalized propynyl structures, at 87% and 74% yields [68].

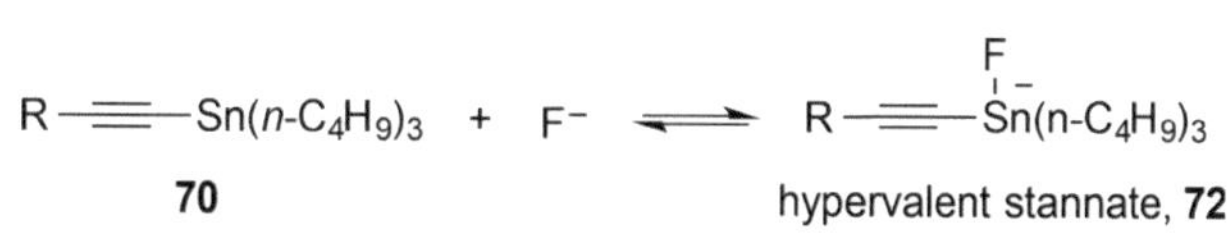

(7)

Synthesis of [^{11}C]iloprost methyl ester: Iloprost (**74**) is a stable prostacyclin (PGI_2) analog specific for the PGI_2 receptor, IP (IP_1), in peripheral systems. It is used as a potential therapeutic agent [69], containing a 1-propynyl moiety on the ω-side chain (see Section 2.4.1.1). The ^{11}C-labeled iloprost methyl ester ([^{11}C]-**74**) was synthesized using [^{11}C]CH_3I and a stannyl precursor (**73**)

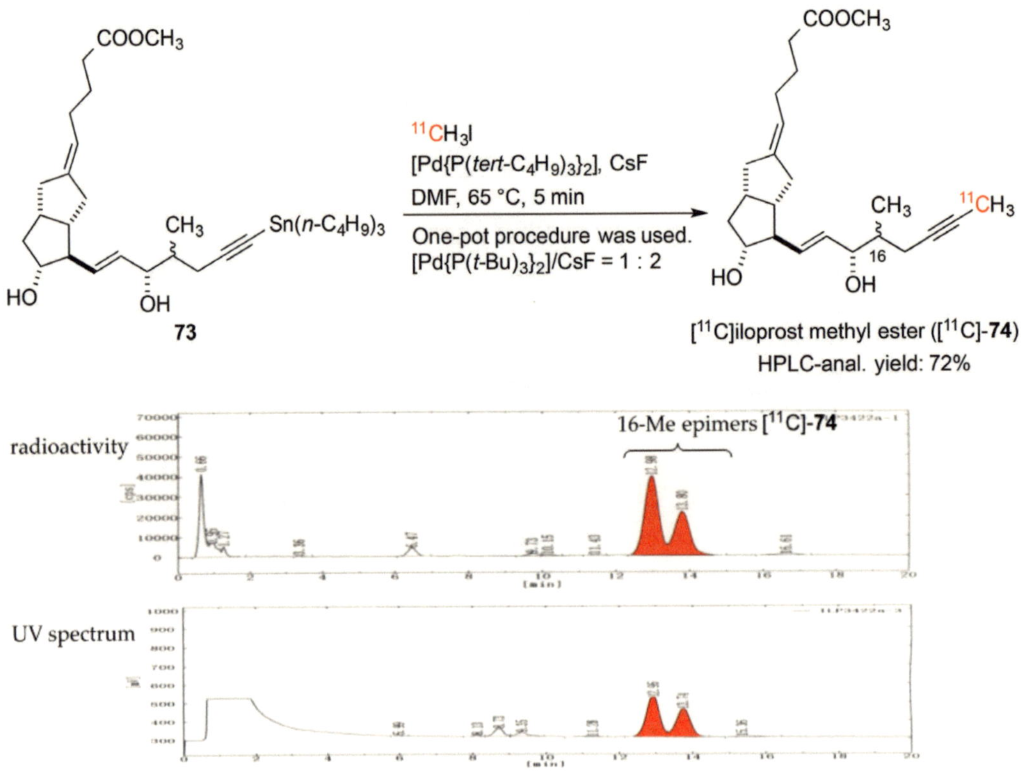

Figure 17. Synthesis of [^{11}C]iloprost methyl ester ([^{11}C]-**74**).

at up to 72% radio-HPLC analytical yield under the conditions noted in **Table 7**, Entry 14 (**Figure 17**) [2]. Thus, we have synthesized both [^{11}C]iloprost methyl ester ([^{11}C]-**74**) and 15*R*-[^{11}C]TIC methyl ester ([^{11}C]-(15*R*)-**26b**) as specific PGI_2 receptor ligands of peripheral and central nervous systems, respectively (**Figure 18**).

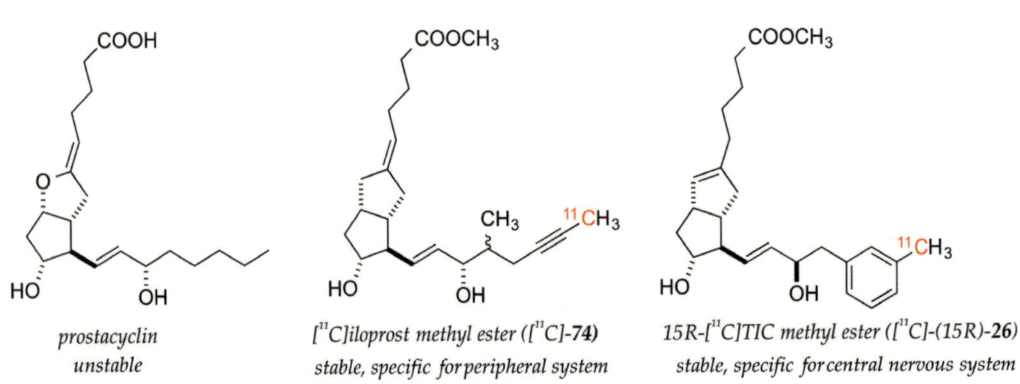

Figure 18. Specific PET probes for prostacyclin receptors in peripheral and central nervous systems.

3.1.5.1. Rapid C-methylation using aryl, alkenyl, and alkyl boronic acid esters as coupling substrates and applications to ^{11}C-labelings

Organoboranes are less toxic than organostannanes. We intended to elaborate the rapid *C*-methylation based on Suzuki-Miyaura cross-coupling reaction [70] as a complementary method to the Stille reaction. The Merck group reported the syntheses of [^{11}C]-**46** derivatives using [^{11}C]CH_3I and an excess amount of phenylboronic acid ester in the presence of [$PdCl_2$(dppf)] and K_3PO_4 in DMF under microwave heating conditions at high temperatures [71]. We intended to establish a more efficient and practical method using a Pd^0-bulky phosphine complex without microwave heating in view of the careful treatment needed for a radiolabeled compound [72].

3.1.5.1.1. Pd^0-Mediated rapid C-methylation of aryl- or alkenylboronic acid ester: Rapid couplings between sp^2(aryl, vinyl)-sp^3 hybridized carbons

We set up a reaction using phenylboronic acid pinacol ester (**75**) ([CH_3I]/[**75**] = 1:40) with a short reaction time fixed at 5 min [72]. The reaction under general Suzuki-Miyaura cross-coupling conditions [70] did not give satisfactory results (24–39% yields; **Table 8A**, Entries 1–3). Therefore, we used the Pd^0 complex coordinated to a bulky phosphine ligand inspired by our earlier success with Pd^0-mediated rapid *C*-[^{11}C]methylations using organostannanes [40]. The reaction rate was dramatically accelerated by the tri-*o*-tolylphosphine Pd complex in DMF in the presence of K_2CO_3, K_2CO_3/H_2O, Cs_2CO_3, or K_3PO_4 to afford the desired toluenes at 87–94% yields (**Table 8A**, Entries 4–7). The rapid coupling reaction of CH_3I and (**75**) probably proceeds via several steps as follows (Eqs. (1), (8)–(11)) [72]:

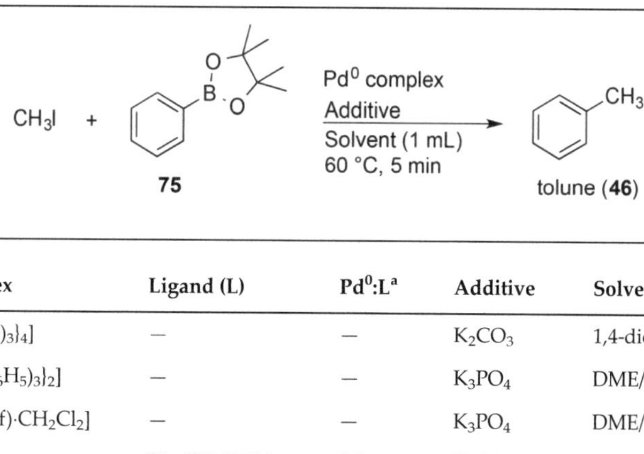

Entry.	Pd^0 complex	Ligand (L)	Pd^0:L[a]	Additive	Solvent	Yield, %
1	[Pd{P(C_6H_5)$_3$}$_4$]	—	—	K_2CO_3	1,4-dioxane	39
2	[$PdCl_2${P(C_6H_5)$_3$}$_2$]	—	—	K_3PO_4	DME/H_2O	24
3	[$PdCl_2$(dppf)·CH_2Cl_2]	—	—	K_3PO_4	DME/H_2O	28
4	[Pd_2(dba)$_3$]	P(*o*-$CH_3C_6H_4$)$_3$	1:2	K_2CO_3	DMF	91
5	[Pd_2(dba)$_3$]	P(*o*-$CH_3C_6H_4$)$_3$	1:2	K_2CO_3	DMF/H_2O	94
6	[Pd_2(dba)$_3$]	P(*o*-$CH_3C_6H_4$)$_3$	1:2	Cs_2CO_3	DMF	92
7	[Pd_2(dba)$_3$]	P(*o*-$CH_3C_6H_4$)$_3$	1:2	K_3PO_4	DMF	87

[a]Molar ratio.

Table 8A. An optimization of conditions for rapid cross-coupling of methyl iodide with phenylboronic acid pinacol ester.

$$CH_3I + [Pd\{P(o\text{-}CH_3C_6H_4)_3\}_2] \longrightarrow \underset{\mathbf{47}}{[Pd(CH_3)I\{P(o\text{-}CH_3C_6H_4)_3\}]} + P(o\text{-}CH_3C_6H_4)_3 \quad (1A)$$

$$\underset{\mathbf{75}}{C_6H_5Bpin} + K_2CO_3 \text{ (or KOH)} \rightleftharpoons \underset{\mathbf{76}}{K^+[(C_6H_5)B(pin)(OZ)]^-} \quad (8)$$

$$\mathbf{47} + \mathbf{76} \longrightarrow \underset{\mathbf{50}}{[Pd(CH_3)(C_6H_5)\{P(o\text{-}CH_3C_6H_4)_3\}]} + \underset{\mathbf{77}}{K^+[B(pin)(OZ)(I)]^-} \quad (9)$$

$$\mathbf{77} + K_2CO_3 \text{ (or KOH)} \longrightarrow \underset{\mathbf{78}}{K^+[B(pin)(OZ)_2]^-} + KI \quad (10)$$

$$\mathbf{50} \longrightarrow \underset{\text{toluene } (\mathbf{46})}{CH_3\text{–}C_6H_5} + [Pd\{P(o\text{-}CH_3C_6H_4)_3\}] \quad (11)$$

Z = COOK or H

(i) oxidative addition to afford the methyl-Pd^{II} iodide (**47**) (Eq. (1)); (ii) the formation of the boronate complex (**76**) (Eq. (8)); (iii) the substitution reaction of (**47**) with (**76**) to afford $[Pd^{II}$(methyl)(phenyl)] complex (**50**) and unstable borate $K^+[B(pin)(OZ)(I)]^-$ (**77**; Z = COOK or H) (Eq. (9)); (iv) transformation of (**77**) to the more stable borate, $K^+[B(pin)(OZ)_2]^-$ (**78**, Eq. (10)); (v) the reductive elimination reaction of (**50**) to afford (**46**) (Eq. (11)).

The conditions were applicable to various aryl, alkenyl, and heteroaromatic-ring substituted boronic acid esters (**Table 8B**) [72]. Thus, such reactions smoothly proceeded under the following conditions: CH_3I/borane/$[Pd_2(dba)_3]$/$P(o\text{-}CH_3C_6H_4)_3$/K_2CO_3 (1:40:0.5:2:2) in DMF at 60°C for 5 min, affording the corresponding methylation products at 80–99% yields, in which the *E* and *Z* olefinic stereoisomers afforded the corresponding methylated products with the retention of stereochemistry. The boronic acid esters containing a more lipophilic ester moiety showed the same reactivity as the pinacol ester, thus making the purification of the labeled probe easier [72].

The conditions (**Table 8A**, Entry 4) were applied to the synthesis of $[^{11}C]$*p*-xylene ($[^{11}C]$-**80**) (**Figure 19**), using pinacol tolylboronate (**79**) at 96% radio-HPLC analytical yield [72].

Synthesis of $[^{11}C]$celecoxib and its metabolite, $[^{11}C]$SC-62807*:* Celecoxib (**82**) is a selective cyclooxygenase (COX)-2 inhibitor that shows analgesic and anti-inflammatory effects in patients with rheumatoid arthritis. In humans, celecoxib is extensively metabolized in the liver via successive two-step oxidation, initially to a hydroxymethyl metabolite (SC-60613), and, on further oxidation, to a carboxylic acid metabolite (SC-62807, **83**). The majority of celecoxib is excreted into the bile as (**83**), a substrate of drug transporters for the OATP1B1 and BCRP. $[^{11}C]$Celecoxib ($[^{11}C]$-**82**) was synthesized at 98% radio-HPLC analytical yield from boronic acid pinacol ester (**81**) using $[Pd_2(dba)_3]$/$P(o\text{-}CH_3C_6H_4)_3$/K_2CO_3 (1:4:9 in molar ratio) at 65°C for 4 min in DMF (**Figure 20A** and **B**). Radiochemical yield: 63% (decay-corrected) [47]; total synthesis time: 30 min; radiochemical and chemical purities: >99% and >98%, respectively. $[^{11}C]$SC-62807 ($[^{11}C]$-**83**) was also synthesized at 87% radio-HPLC analytical yield from purified $[^{11}C]$-**82** by rapid oxidation using excess $KMnO_4$ under microwave irradiation conditions

http://dx.doi.org/10.5772/intechopen.72868

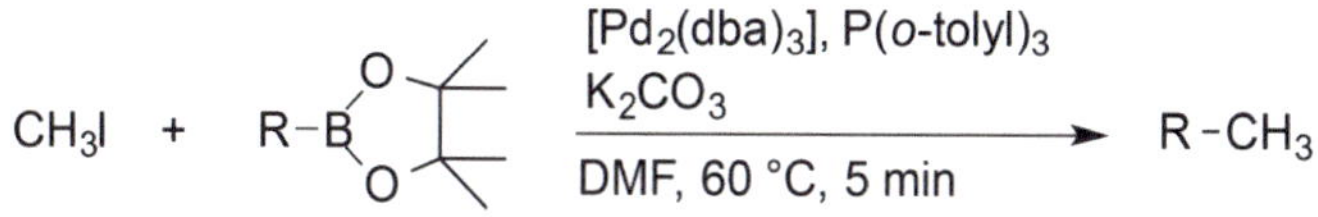

CH_3I/boron/[$Pd_2(dba)_3$]/P(*o*-$CH_3C_6H_4$)$_3$/K_2CO_3 (1:40:0.5:2:2).

Entry	R = arenyl	Yield, %	Entry	R = alkenyl	Yield, %
1	4-$CH_3OC_6H_4$-	98	12	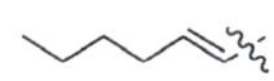	95
2	3-benzyloxyC_6H_4-	95	13		92
3	4-$CH_3C_6H_4$-	99			
4	4-PhC_6H_4-	91	14	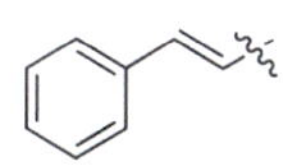	96
5	4-FC_6H_4-	93			
6	4-$CH_3OCOC_6H_4$-	96[a]	15	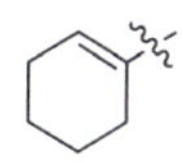	86
7	4-$NO_2C_6H_4$-	99			
8	thiophen-2-yl-	92	16	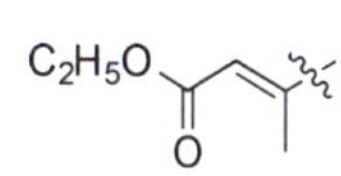	99
9	furan-2-yl-	85			
10	pyridine-4-yl-	80	17	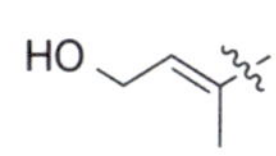	99
11	isoquinolin-4-yl-	85			

[a]Mixed solvent (DMF:H_2O = 9:1) was used.

Table 8B. Rapid cross-coupling of methyl iodide with a wide variety of pinacol arenyl- and alkenylboronates.

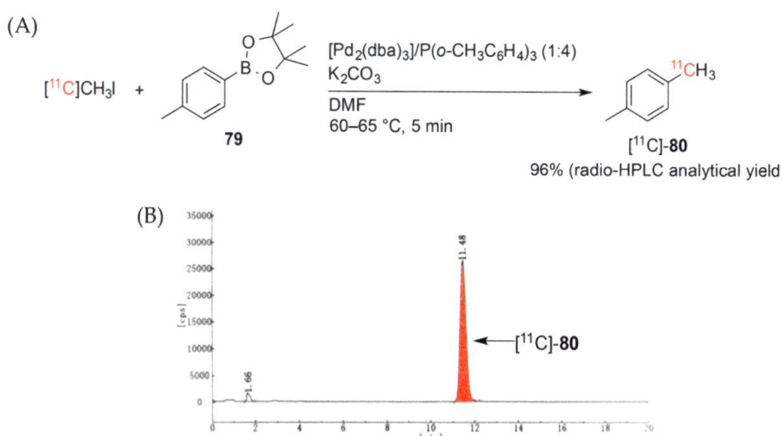

Figure 19. Synthetic scheme of ^{11}C-labeled *p*-xylene ([^{11}C]-**80**) by rapid *C*-[^{11}C]methylation using *p*-tolueneboronic acid pinacol ester (**79**) (A), and radio-HPLC chart in the analysis of the reaction mixture (B).

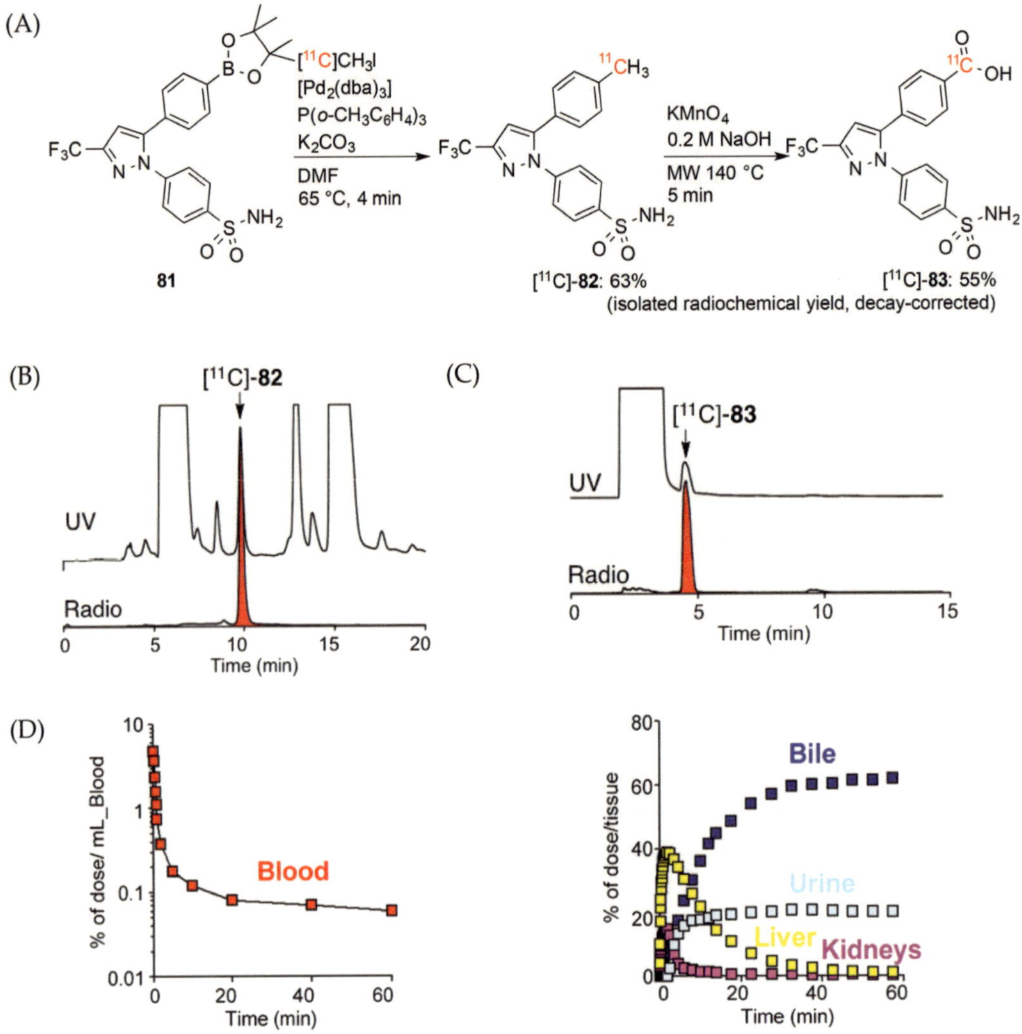

Figure 20. Syntheses of [^{11}C]celecoxib ([^{11}C]-**82**) and [^{11}C]SC-62807 ([^{11}C]-**83**) (A), radio-HPLC analyses of [^{11}C]-**82** (B) and [^{11}C]-**83** (C), and the time profiles of the radio activity of [^{11}C]-**83** in the blood (red point), liver (yellow point), kidney (pink point), bile (blue point), and urinary bladder (light blue point) determined by PET imaging studies and blood sampling over a 60-min period after administration of [^{11}C]-**83** to male Sprague–Dawley rats (D). The peaks at retention times of 9.6 (in B) and 4.4 min (in C) are [^{11}C]-**82** and [^{11}C]-**83**, respectively. UV absorbance: 254 nm.

(**Figure 20A** and **C**), giving [^{11}C]-**83** at 87% radio-HPLC analytical yield. Radiochemical yield: 55% (decay-corrected); synthesis time: 20 min; radiochemical purity: >99%. PET studies in rats and metabolite analyses showed that [^{11}C]-**83** was rapidly excreted via hepatobiliary excretion without further metabolism (**Figure 20D**), and thus has a great potential, as a suitable PET probe, to investigate the hepatobiliary transport process [48].

Synthesis of ^{11}C-incorporated acromelic acid analog: Acromelic acid A isolated from *Clitocybe acromelalga* induces neuropathic pain allodynia. A novel ^{11}C-labeled PET probe, [^{11}C]PSPA-4

([^{11}C]-**85**), was synthesized. The probe has been designed as a simpler acromeric acid structure to prevent the acromelic acid-induced allodynia, and develop novel analgesic drugs and diagnostics in the treatment of neuropathic pain [73]. Thus, the rapid *C*-[^{11}C]methylation of [^{11}C]CH_3I and boronic acid pinacol ester precursor (**84**) by using [Pd_2(dba)$_3$], P(*o*-$CH_3C_6H_4$)$_3$, and K_2CO_3 in DMF followed by rapid deprotection of the TFA-protected amino acid moiety and hydrolysis of methyl esters in a onepot reaction within 5 min (4 and 1 min each) afforded [^{11}C]-**85** at >99% both radiochemical and chemical purities [73]; radiochemical yield: 34–43% (decay-corrected); total radioactivity after purification: 5.0–6.0 GBq (**Scheme 15**); total synthesis time: 30 min.

Synthesis of [^{11}C]all-trans-retinoic acid: Retinoids are a class of chemical compounds including both naturally occurring dietary vitamin A (retinol) metabolites and their synthetic analogs, regulating a wide variety of essential biological processes, such as, vertebrate embryonic morphogenesis and organogenesis, cell growth arrest, differentiation, apoptosis, and homeostasis, and their disorders. All-*trans*-retinoic acid (ATRA, **87**), the most potent biologically active metabolite of retinol, is a ligand for the retinoic acid receptors (RAR-α, $-\beta$, and -γ) and used in the treatment of acute promyelocytic leukemia. [^{11}C]ATRA ([^{11}C]-**87**) was synthesized at 25% radiochemical yield (decay-corrected) from boronic acid ester precursor (**86**) [74] using [Pd_2(dba)$_3$], P(*o*-$CH_3C_6H_4$)$_3$, and Na^+ ascorbate (1:4:9) in DMF/H_2O at 65°C for 4 min, followed by basic hydrolysis of the methyl ester (at 65°C) or ethyl ester (at 100°C) for 2 min; total radioactivity: 1.5 GBq; radiochemical and chemical purities: >99% and 97%, respectively; total synthesis time: 35 min (**Scheme 16**, see also Section 3.3 for [^{11}C]carbonylation).

Synthesis of [^{11}C]dehydropravastatin: The study of transporter functions in the liver is important for drug development and diagnosis of hepatic diseases associated with a particular transporter dysfunction. Organic anion-transporting polypeptides (OATPs) and multidrug resistance-associated protein 2 (MRP2) drug transporters play important roles in the uptake of drugs into the liver and canalicular efflux, respectively [75]. Pravastatin (**88**) [76] has been widely used to lower the cholesterol concentration in blood by inhibiting the action of HMG-CoA reductase in the liver. The uptake of pravastatin into the liver is mediated by the OATP1B1 transporter; the drug is then excreted into the bile by MRP2 without metabolism. Therefore, pravastatin would be a good molecular probe for a PET study to track and evaluate the functions of these two transporters in an *in vivo* system [77]. We designed 2′,3′-dehydropravastatin (**89**) as it has easier synthetic accessibility and is a suitable target for our new synthetic

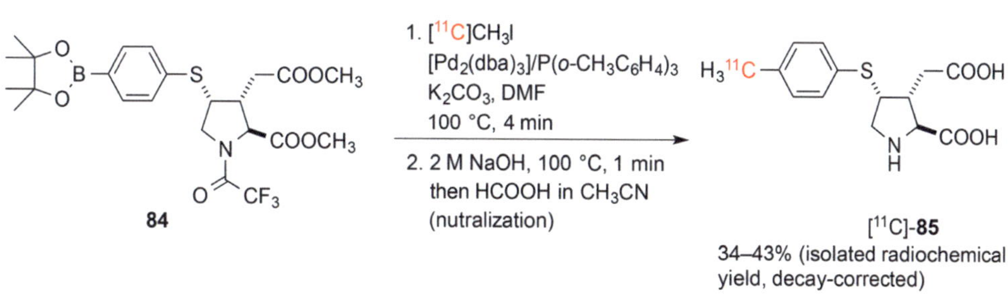

Scheme 15. Synthesis of ^{11}C-labeled acromelic acid analog [^{11}C]-**85**.

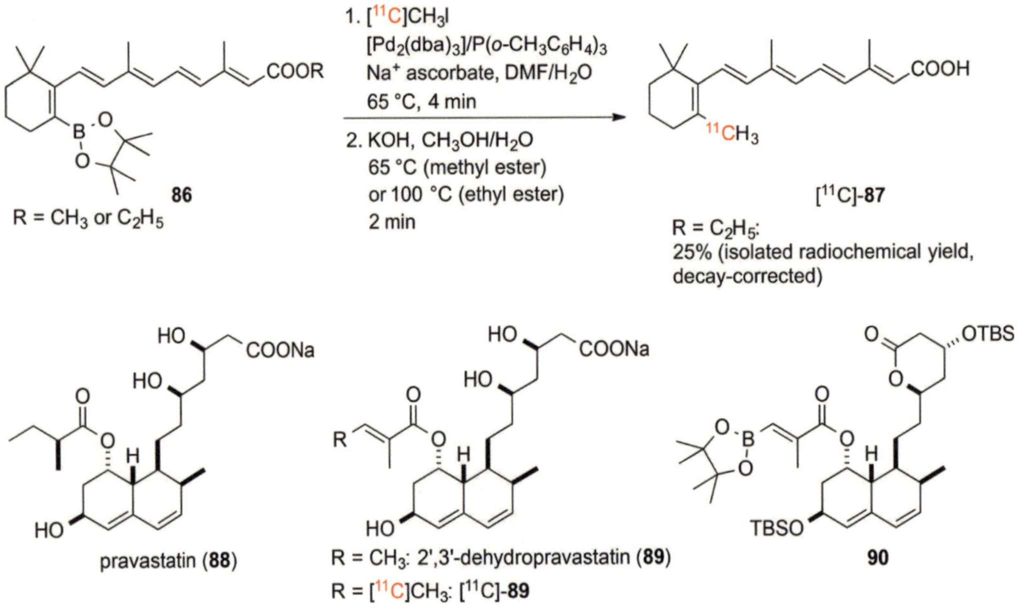

Scheme 16. Synthesis of ^{11}C-labeled all-*trans*-retinoic acid ([^{11}C]ATRA, [^{11}C]-**87**).

methodology. [^{11}C]-**89** was synthesized from protected boronate (**90**) using [$Pd_2(dba)_3$], P(*o*-$CH_3C_6H_4)_3$, and K_2CO_3 (1:3:4) at 65°C for 5 min in THF to afford a mixture including the desired ^{11}C-incorporated compound [77]. Successive treatments with TBAF in THF and an aqueous NaOH solution, and then the isolation by HPLC afforded [^{11}C]-**89** at radiochemical yield (decay-corrected), 20%; total radioactivity, 1.2 GBq; radiochemical purity, >99%; chemical purity, 98%; total synthesis time: 34 min. The PET studies on rats showed that [^{11}C]-**89** has great potential to be a suitable PET probe to evaluate the functions of OATPs and MRP2 in hepatobiliary excretion (**Figure 21**) [77, 78].

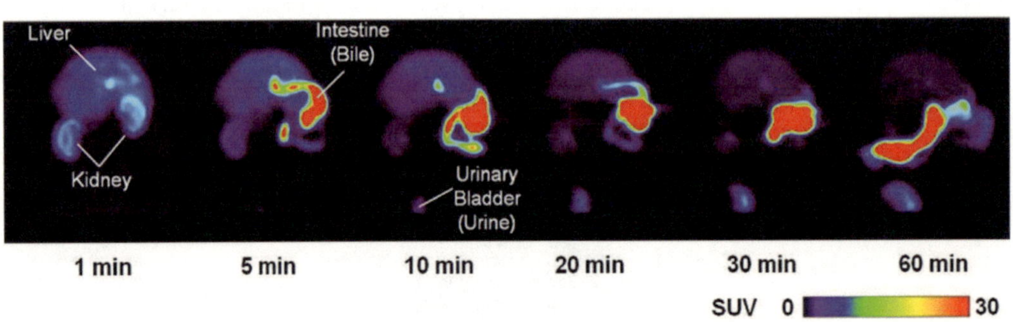

Figure 21. Color-coded PET images of the abdominal regions of rats after administration of [^{11}C]-**89**; radioactivity was identifiable in the liver, kidneys, and localized mainly in the intestine (via the bile excreted into the intestine) and urinary bladder (via the urine excreted into the urinary bladder). The radioactivity in the organs was modulated after co-administration with a typical OATP inhibitor or by using MRP2 hereditary-deficient rats.

http://dx.doi.org/10.5772/intechopen.72868

3.1.5.1.2. Rapid C-methylation reaction of benzyl or cinnamyl boronic acid ester (rapid coupling of sp^3–sp^3 hybridized carbons)

The transition metal-mediated cross-coupling reactions using alkylboronate produce undesired alkanes or alkenes induced by ligand scrambling or β-elimination reaction [79–81], causing safety and environmental problems during the synthesis of PET probes. Thus, the rapid cross-coupling reactions of CH_3I with excess benzyl- and cinnamylboronic acid pinacol ester lacking hydrogen on neighboring sp^3-carbon were studied [82]. We particularly focused on using a combination of bulky triaryl- or trialkylphosphines/carbonate or fluoride as the base in DMF, in accordance with our previous studies (**Table 9**). Thus, the reactions using CH_3I/**91**/[Pd{P(*tert*-$C_4H_9)_3\}_2$]/CsF (1:40:1:10) in 90:10 DMF/H_2O (*v*/*v*) at 80°C for 5 min and CH_3I/**91**/[$Pd_2(dba)_3$]/P(*o*-$CH_3C_6H_4)_3/K_2CO_3$ (1:40:0.5:2:2) in 90:10 DMF/H_2O (*v*/*v*) at 120°C for 5 min afforded ethylbenzene (**92**) at 88% and 43% yields, respectively, as the optimized conditions for the use of the corresponding bulky phosphines (**Table 9**, Entries 7 and 2, respectively) [82]. Interestingly, the yield obtained using the former reaction conditions was temperature dependent, resulting in a bell-shaped curve (**Figure 22**). The use of an excess amount of substrate (**91**) (200 equiv. for methyl iodide) gave positive results for the former reaction leading to quantitative yield of the product (**Table 9**, Entry 7), thus agreeing well with an actual PET study using largely excess trapping substrate (see Section 3).

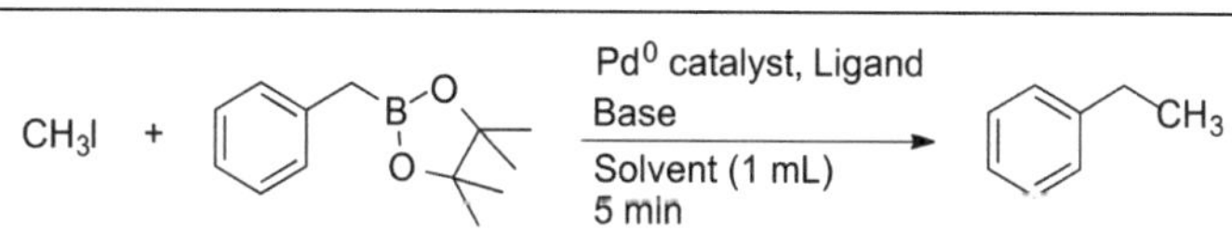

CH_3I/91/Pd^0/ligand/Base (1:40:1:2:2–10).

Entry	Pd[0a]	Base	Solvent	Yield, %			
		(equiv)		60 °C	80 °C	100°C	120°C
1	A	K_2CO_3 (2)	DMF	6	–	18	–
2	A	K_2CO_3 (2)	DMF/H_2O	20	32 (27)[c]	39 (45)[c]	43 (37)[c]
3	A	CsF (10)	DMF/H_2O	34	47 (36)[c]	36	40
4	B	K_2CO_3 (2)	DMF/H_2O	–	40 (70)[c]	32	25
5	B	CsF (10)	DMF	42	58	–	–
6	B	KF (10)	DMF/H_2O	–	62	61	–
7	B	CsF (10)	DMF/H_2O	54	**88 (100)**[c]	62	6–59
8	B[b]	CsF (10)	DMF/H_2O	–	61	–	–

[a]Pd^0/ligand:A, [$Pd_2(dba)_3$]/P(*o*-$CH_3C_6H_5)_3$; B, [Pd{P(*tert*-$C_4H_9)_3\}_2$].
[b]Excess P(*tert*-$C_4H_9)_3$ (16 equiv).
[c]Reaction using a five-fold excess of boronate substrate (200 equiv).

Table 9. Rapid cross-coupling of methyl iodide and benzylboronic acid pinacol ester (**91**).

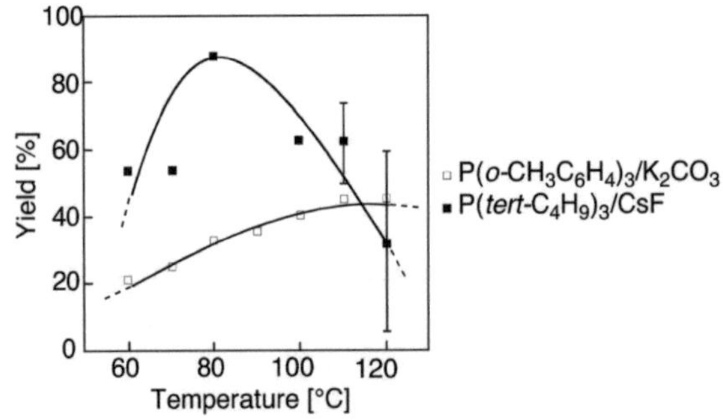

Figure 22. The yields of ethylbenzene by the reactions at various temperatures using CH_3I/**91**/[$Pd_2(dba)_3$]/P(*o*-$CH_3C_6H_4)_3/K_2CO_3$ (1:40:0.5:2:2) (white square) and CH_3I/**91**/[Pd{P(*tert*-$C_4H_9)_3\}_2$]/CsF (1:40:1:10) (black square) in 90:10 DMF/H_2O (*v*/*v*) for 5 min.

The use of a bulky trialkylphosphine (cone angle = 182° for P(*tert*-$C_4H_9)_3$) with its much higher basicity (pK_a = 11.4 for P(*tert*-$C_4H_9)_3$) [83] relative to an arylphosphine (pK_a = 3.8 for P(*o*-$CH_3C_6H_4)_3$) [83] could be important to promote the rapid C-methylation. As shown in **Figure 23**, P(*tert*-$C_4H_9)_3$ with a strong σ electron-donating ability can generate a coordinatively unsaturated tricoordinated Pd^{II} complex, CH_3Pd^{II}(L)I (**94**), by oxidative addition (Eq. (12)), which could be in equilibrium with the highly polarized (or ionized) Pd(II) complex $[CH_3Pd^{II}(L)]^+I^-$ (**94′**) owing to strong electron donation from the phosphine ligand. The resulting complex (**94**) (or **94′**) readily undergoes an/a iodide/benzyl group exchange upon reaction with fluoro benzylboronate (**95**) formed by coordination of fluoride to boron (Eqs. (13) and (14)), affording a tricoordinated complex (**96**) and Cs salt (**97**). The former undergoes

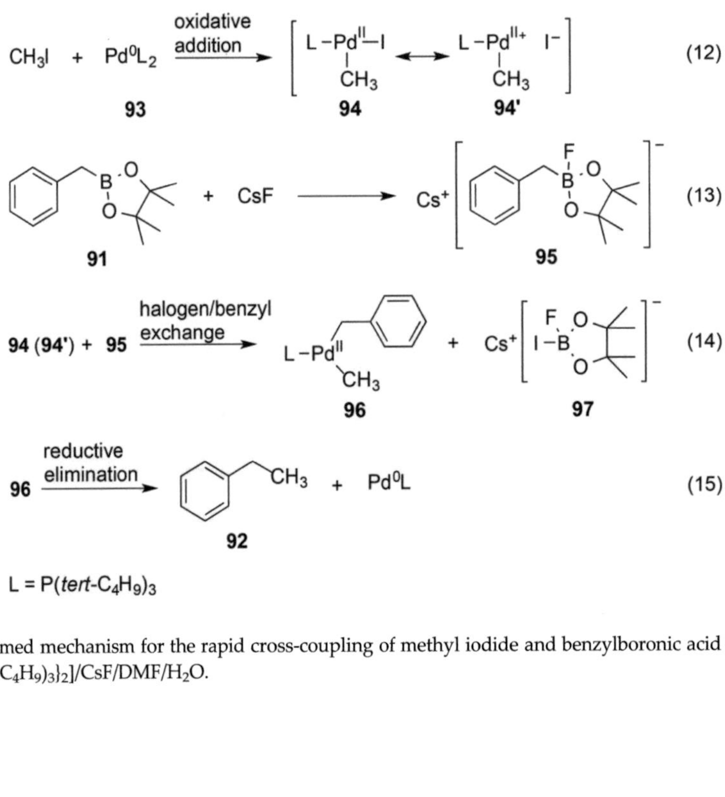

Figure 23. Assumed mechanism for the rapid cross-coupling of methyl iodide and benzylboronic acid pinacol ester (**91**) using [Pd{P(*tert*-$C_4H_9)_3\}_2$]/CsF/DMF/H_2O.

reductive elimination (Eq. (15)) to afford (**92**). Unsaturated tricoordinated Pd^{II} complex (**96**) could be directly responsible for the reductive elimination.

Optimized reaction conditions were applied to various types of benzyl- and cinnamylboronic acid pinacol esters, in which electron-donating groups such as *p*-methoxy, 3,4-dimethoxy, *p*-amide, and *p*-methyl groups or electron-withdrawing groups such as *p*-chloro, *p*-fluoro, *m*-ethyl ester, and *m*-trifluoromethyl groups are substituted on the phenyl rings, resulting in 80–100% yields (**Table 10**, Entries 4, 5, 7, 9, 10, and 12–14), while the boronic acid esters possessing *o*-fluoro or *p*-trifluoro substituents on the phenyl ring were less reactive (**Table 10**, Entries 8 and 11) [82].

The utility of a rapid reaction was demonstrated by reacting [^{11}C]CH_3I and boronic acid ester (**98c**) in the presence of [Pd{P(*tert*-$C_4H_9)_3$}$_2$]/CsF (1:10) in 90:10 DMF/H_2O (*v*/*v*) at 90°C for 5 min to afford *N*-(4-[^{11}C]ethylphenyl)propionamide ([^{11}C]-**99c**) in 90% radio-HPLC analytical yield (**Figure 24A** and **B**). Radiochemical yield (decay-corrected): 49%; total radioactivity: 5.9 GBq; radiochemical purity: >99%; chemical purity: 95%; total synthesis time: 32 min [82].

Here, we compared the effects of TlOH and conventional bases, K_2CO_3 and CsF, for the Pd^0-mediated rapid cross-coupling based on the use of CH_3I with organoboranes (**Table 11**) [84]. It should be noted that the rapid *C*-methylation between CH_3I and organoborane reagents such as phenyl-(**75**), (*Z*)-4-benzyloxy-2-butenyl-(**100**), and benzylboronic acid pinacol esters (**91**) did not proceed at all for 5 min under the reported conditions: CH_3I/borane/[Pd{P($C_6H_5)_3$}$_4$]/TlOH [85] (1:40:1:3) in THF/H_2O at 25°C. Even the use of bulky phosphines at elevated temperatures gave much lower yield than our findings. Thus, it was concluded that the use of TlOH is not applicable to the cross-coupling reaction of CH_3I and various types of organoborons substrates, reconfirming that the use of the milder and non-toxic bases, K_2CO_3 or CsF, are most appropriate for rapid *C*-[^{11}C]methylation [84].

3.2. Rapid *C*-[^{18}F]fluoromethylation

C-[^{18}F]Fluoromethylation using [^{18}F]FCH_2Br or [^{18}F]FCH_2I is related to the *C*-[^{11}C]methylation described in Section 3.1. The advantages of ^{18}F–labeling are as follows: (1) longer *in vivo* monitoring as compared to ^{11}C, (2) delivery of ^{18}F–labeled probes to distant PET centers and clinics, (3) the possible use of multi-reactions after labeling, and (4) the possible use in click chemistry as a prosthetic group for the labeling of peptides, nucleic acids, sugars, etc.

According to the reaction conditions established using FCH_2I [72, 86], we set up a reaction using ca. 0.5 GBq of [^{18}F]FCH_2X (X = Br or I) and a 4-(4,4,5,5-tetramethyl-1,3,2-dioxaborolan-2-yl)benzoic acid methyl ester (**101**) (**Figure 25**). The labeling using [^{18}F]FCH_2I and (**101**) in the presence of [Pd_2(dba)$_3$]/P(*o*-$CH_3C_6H_4)_3$ (1:6) and K_2CO_3 in DMF for 5 min at 65°C afforded the desired *p*-([^{18}F]fluoromethyl)benzoic acid methyl ester ([^{18}F]-**102**) at 23% radio-HPLC analytical yield. After a broader investigation, we concluded that the *C*-[^{18}F]fluoromethylation using [^{18}F]FCH_2Br would be more practical than that using [^{18}F]FCH_2I owing to the instability of the latter under light-exposed conditions. Thus, the reaction at 120°C for 15 min in DMPU/H_2O (9:1) afforded [^{18}F]-**102** at 60–80% yields [86]. The 15-min reaction thus obtained agrees with the requirements of ^{18}F–incorporated PET-probe synthesis because of the longer half-life

Entry	Boronic acid ester	Yield, %				
		60 °C	70°C	80 °C	100°C	120°C
1	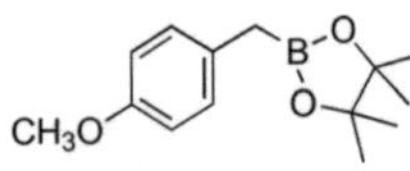 98a	98	—	100	—	—
2	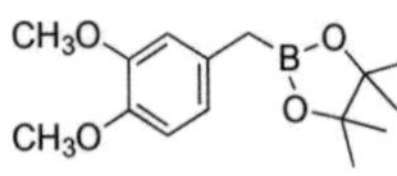 98b	77	—	80	—	81
3	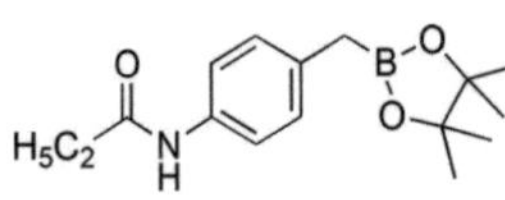 98c	91	—	100	—	—
4	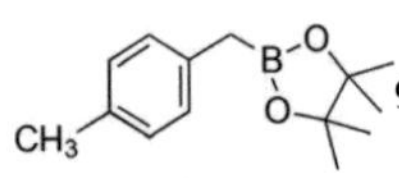 98d	65	—	91	—	—
5	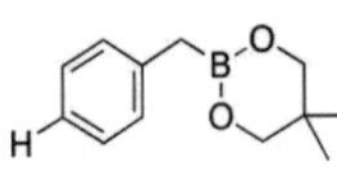 98e	—	75	72	—	—
6	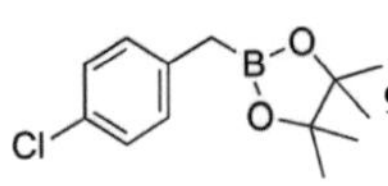 98f	80	—	84	—	100
7	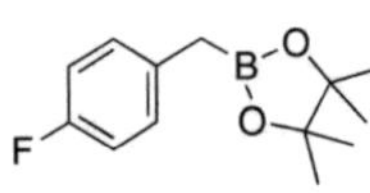 98 g	—	—	82	—	68
8	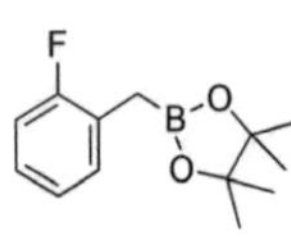 98 h	6	—	39	34	35
9	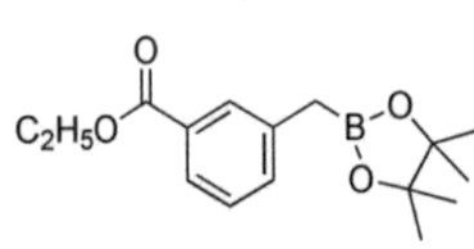 98i	69	—	98	—	—
10	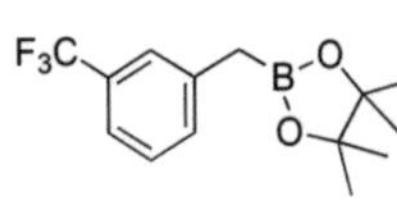 98j	87	—	91	—	—
11	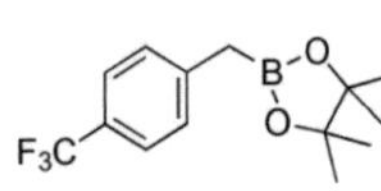 98 k	—	—	19	—	21
12	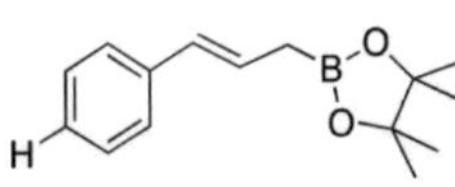 98 l	—	—	43	100	33
13	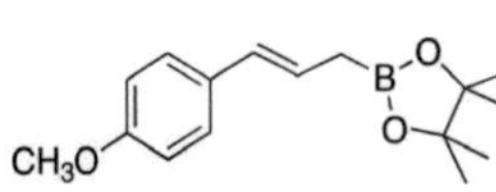 98 m	30	37	87	59	58
14	$(CH_3BO)_3$98n	—	—	100	—	—

Table 10. Rapid cross-couplings of methyl iodide and benzyl-(**98a–98d, 98f–98 k**) and cinnamylboronic acid pinacol esters (**98 l–98 m**), benzylboronic acid neopentyl glycol ester (**98e**), and trimethylboroxine (**98n**) and the corresponding coupling products (**99a–n**). Reaction under the conditions of CH_3I/**98**/[Pd{P(*tert*-C_4H_9)$_3$}$_2$]/CsF (1:40:1:10) in 90:10 DMF/H_2O (*v*/*v*) for 5 min.

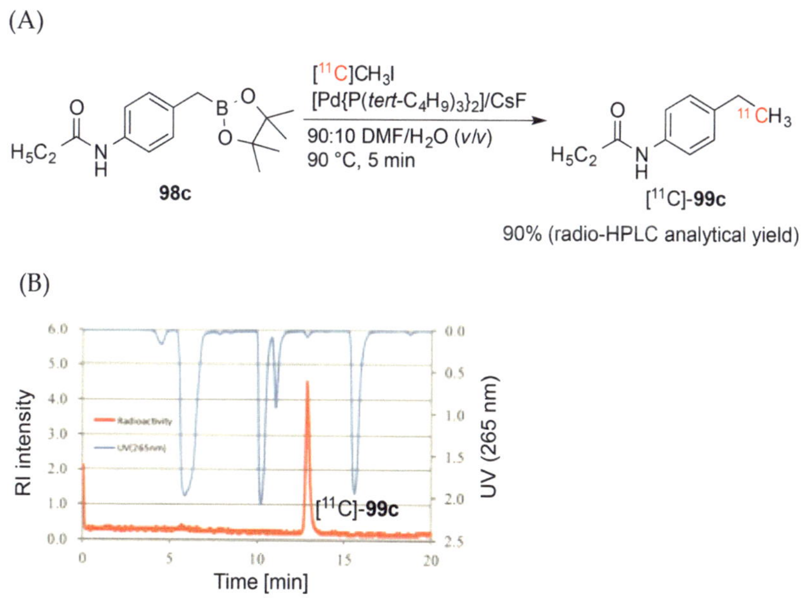

Figure 24. Synthesis of *N*-(4-[^{11}C]ethylphenyl) propionamide ([^{11}C]-**99c**) (A) and radio-HPLC analysis (B).

(109.8 min) of the ^{18}F radionuclide compared to ^{11}C (20.4 min) [86]. The rapid reaction procedure afforded 6-([^{18}F]fluoromethyl)quinoline ([^{18}F]-**103**) at 69% radio-HPLC analytical yield. Moreover, [^{18}F]fluoromethylated compounds [^{18}F]-**104** and [^{18}F]-**105** were obtained at 45% and 40% radio-HPLC analytical yields [86]. The corresponding acid of *p*-[^{18}F]fluoromethyl benzoate [^{18}F]-**102** could be used as a prosthetic group for ^{18}F–labeled biomarkers with higher molecular weights such as oligonucleotides, peptides, and proteins.

3.3. Rapid *C*-[^{11}C]carbonylation under ambient temperature and atmospheric pressure using conventional helium gas as the [^{11}C]CO carrier

The low solubility of CO in most common solvents has hampered the Pd^{0}-mediated rapid cross-coupling reaction. Therefore, most of such reactions have been conducted with the use of special equipment such as high-pressure vessels to facilitate the incorporation of [^{11}C]CO into the reaction [87]. Chemical [^{11}C]CO fixation techniques have recently been developed as an alternative with the aim of increasing the concentration of [^{11}C]CO in solution [88]. Pd^{0}-mediated [^{11}C]carbonylation has enabled the execution of the reaction under atmospheric pressure using xenon as a carrier gas, an aryl iodide as a substrate, and an amine and alcohol as trapping nucleophiles for the synthesis of ^{11}C-incorporated amide, urea, and carbamates [89]. Yamamoto et al. elaborated the methoxycarbonylation of an arylboronic acid or esters (2,2-dimethylpropane-1,3-diol ester and pinacol ester), which proceeds under atmospheric pressure at RT for 3–20 hrs, using a catalytic amount of Pd^{II} acetate/triphenylphoshine [Pd(OAc)$_2$]/PPh$_3$ in the presence of *p*-benzoquinone (pbq) in CH_3OH, represented by the following mechanism [90]: (1) the transmetalation of the aryl group from boron to palladium (Eq. (16)); then, insertion of CO into the Pd-C(sp^2) bond (Eq. (17)); (3) reaction with alcohol to

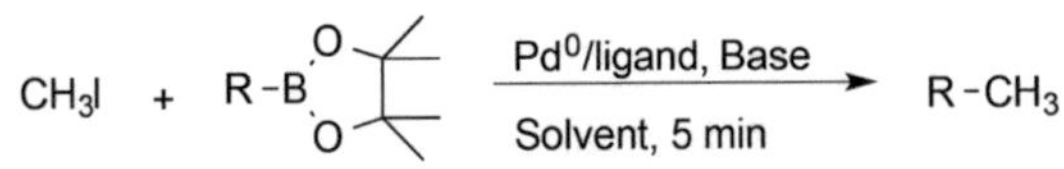

CH_3I/Boron/Pd^0/PR_3/Base (1:40:1:2-4:2-3 in molar ratio)

Entry	Boronic acid ester	Pd^0/ligand	Base	Solvent	Yield, %		
					25°C	60 °C	80 °C
1	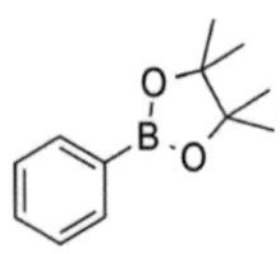 75	$[Pd(PPh_3)_4]$	TlOH	THF/H_2O	0	39	—
2		$[Pd_2(dba)_3]$ /$P(o\text{-}CH_3C_6H_4)_3$	TlOH	THF/H_2O	60	65	—
3		$[Pd_2(dba)_3]$ /$P(o\text{-}CH_3C_6H_4)_3$	KOH	THF/H_2O	—	93	—
4		$[Pd_2(dba)_3]$ /$P(o\text{-}CH_3C_6H_4)_3$	K_2CO_3	DMF/H_2O	—	94	—
5	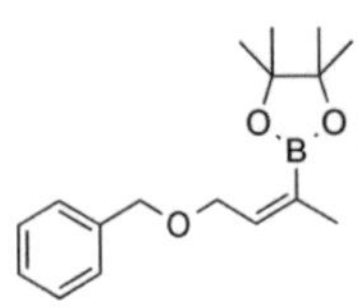 100	$[Pd(PPh_3)_4]$	TlOH	THF/H_2O	0	38	—
6		$[Pd_2(dba)_3]$ /$P(o\text{-}CH_3C_6H_4)_3$	TlOH	THF/H_2O	26	64	—
7		$[Pd_2(dba)_3]$ /$P(o\text{-}CH_3C_6H_4)_3$	KOH	THF/H_2O	—	69	—
8		$[Pd_2(dba)_3]$ /$P(o\text{-}CH_3C_6H_4)_3$	K_2CO_3	DMF/H_2O	—	95	—
9	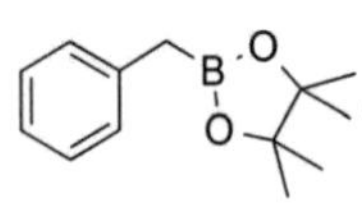 91	$[Pd(PPh_3)_4]$	TlOH	THF/H_2O	0	—	14
10		$[Pd_2(dba)_3]$ /$P(o\text{-}CH_3C_6H_4)_3$	TlOH	DMF/H_2O	—	48	41
11		$[Pd_2(dba)_3]$ /$P(o\text{-}CH_3C_6H_4)_3$	KOH	DMF/H_2O	—	64	78
12		$[Pd_2(dba)_3]$ /$P(o\text{-}CH_3C_6H_4)_3$	CsF	DMF/H_2O	—	54	88

Table 11. Effects of bases for rapid trapping of methyl iodide with an excess amount of various boronic acid esters: focus on the use of TlOH.

give a desired product (Eq. (18)); (4) oxidation of the Pd(0) species with pbq to give Pd(II) species, establishing a catalytic cycle (Eq. (19)).

$$ArB(OR)_2 + LnPd(II)X_2 \longrightarrow XB(OR)_2 + LnPd(II)X\text{-}Ar \quad (16)$$

(17)

(18)

$$LnPd(0) \xrightarrow[2X^-]{pbq} LnPd(II)X_2 \quad (19)$$

$X = CH_3COO$

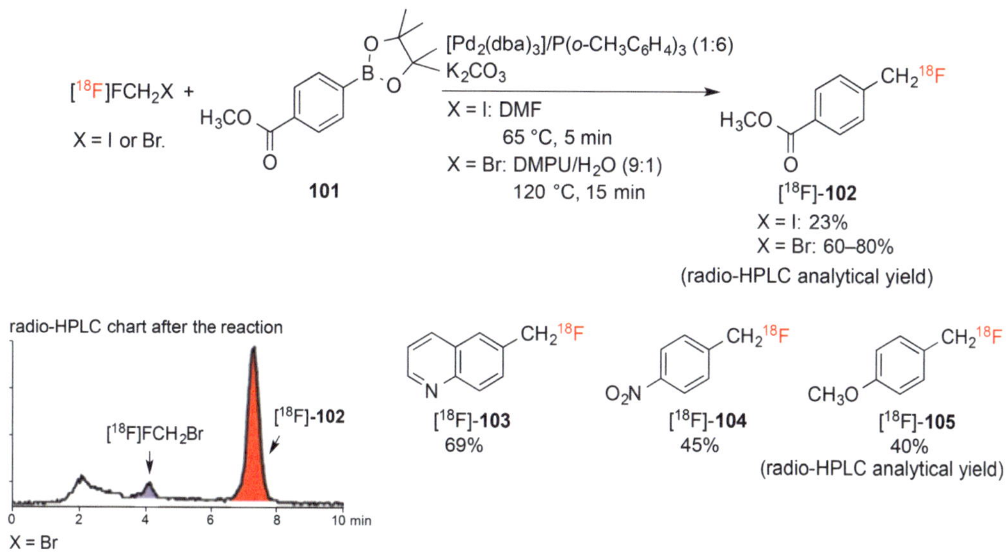

Figure 25. Rapid *C*-[^{18}F]fluoromethylation of an arylboronic acid pinacol ester (**101**) with [^{18}F]FCH_2X (X = I or Br).

Our interest had been focused on the [^{11}C]CO fixation method and Yamamoto's methoxycarbonylation conditions. The synthesis of [^{11}C]benzoic acid methyl ester ([^{11}C]-**107**) was conducted using the reaction of phenylboronic acid 2,2-dimethlpropane-1,3-diol ester (**106**) and [^{11}C]CO as a model reaction (**Table 12**) [91]. However, such conditions gave the desired product at lesser than 3% decay-corrected radiochemical yield (RCY) at RT for 5 min (Entries 1–7). Here, we have *surprisingly* found that rapid methoxy[^{11}C]carbonylation can be facilitated simply by adding DMF in these conditions at RT for 5 min, giving the product [^{11}C]-**107** at 25% RCY (Entry 8). DMSO or THF were ineffective for this purpose. The improved reaction conditions (the use of a twofold excess of the solvent and reagents and the ^{11}CO trapping at lower temperature (−15°C)) were highly favorable to give [^{11}C]-**107** (66% RCY, Entry 9) and [^{11}C]methyl 4-((5,5,8,8-tetramethyl-5,6,7,8-tetrahydronaphthalen-2-yl)carbamoyl)benzoate ([^{11}C]-**108**) (65% RCY). One-pot synthesis of [^{11}C]carboxylic acid was also established by the combination of [^{11}C]carbomethoxylation and rapid hydrolysis (80°C, 2 min) with aqueous KOH [91]. Artificial retinoid Am80 (Tamibarotene, **110**) is an active agent for acute promyelocytic leukemia (APL) patients who have relapsed from ATRA-induced complete remission. It binds tightly to RAR-α and does not bind to the RAR-γ receptor (see also Section 3.1.5.1.1). The above improved conditions were applied to the synthesis of [^{11}C]Am80 ([^{11}C]-**110**) (**Scheme 17**) [91]. The synthesis was conducted in a one-pot manner without isolation of [^{11}C]-**108**. Thus, after methoxy[^{11}C]carbonylation of boronate precursor (**109**), [^{11}C]-**108** underwent rapid hydrolysis with aqueous KOH at 80°C over 2 min to afford the desired acid, [^{11}C]-**110,** at 48% RCY based on [^{11}C]CO with 99% radiochemical purity. The total synthesis time for formulation of [^{11}C]-**110** was 36 min. The isolated radioactivity was 1.4 GBq, and the molar activity was 44 GBq/μmol. This new method is also used to achieve the direct syntheses of [^{11}C]aspirin ([^{11}C]-**111**) using water or tetramethylammonium hydroxide as the hydroxyl source and [^{11}C] salicylamide ([^{11}C]-**112**) and [^{11}C]nicotinamide ([^{11}C]-**113**) with an aqueous ammonia solution as the nucleophile instead of CH_3OH in DMF [92]. This simple [^{11}C]carbonylation method,

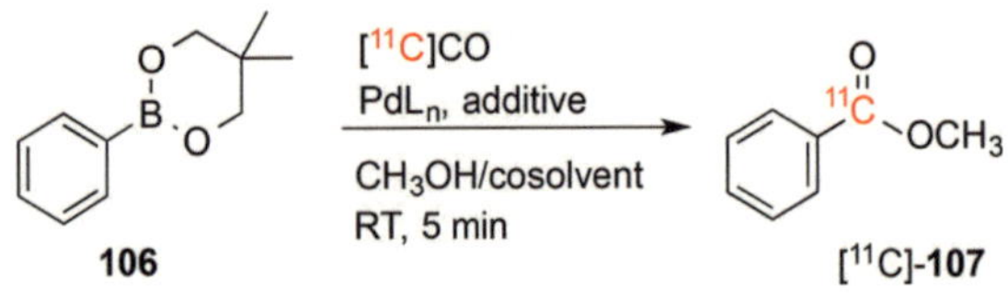

Entry	Conditions	Solvent	Trap. Temp.	Reaction Temp.	RCY, %
1	1. CuCl, K[Tp*], THF; 2. [Pd(PPh_3)$_4$], PPh_3	THF-CH_3OH-DMF (10:1:4)	RT	100°C	<1
2	1. CuCl, K[Tp*], THF; 2. [Pd(OAc)$_2$], P(o-tolyl)$_3$	THF-CH_3OH-DMF (10:1:4)	RT	100°C	<1
3	1. CuCl, K[Tp*], THF; 2. [Pd(PPh_3)$_4$], PPh_3	THF-CH_3OH-DMF (10:1:4)	RT	RT	<1
4	1. CuCl, K[Tp*], THF; 2. [Pd(OAc)$_2$], PPh_3, pbq	THF-CH_3OH-DMF (10:1:4)	RT	RT	3
5	[Pd(OAc)$_2$], PPh_3, pbq	CH_3OH	RT	RT	2
6	[Pd(OAc)$_2$], PPh_3, pbq	CH_3OH-DMSO (1:1)	RT	RT	<1
7	[Pd(OAc)$_2$], PPh_3, pbq	CH_3OH-THF (1:1)	RT	RT	<1
8	[Pd(OAc)$_2$], PPh_3, pbq	CH_3OH-DMF (1:1)	RT	RT	25
9	[Pd(OAc)$_2$], PPh_3, pbq	CH_3OH-DMF (1:1)	−15 °C	RT	26, 66[a]
10	[Pd(OAc)$_2$], PPh_3	CH_3OH-DMF (1:1)	−15 °C	RT	20
11	[Pd(OAc)$_2$], PPh_3, CsF	CH_3OH-DMF (1:1)	−15 °C	RT	<5
	[Pd_2(dba)$_3$], P(o-tolyl)$_3$	CH_3OH-DMF (1:1)	−15 °C	RT	4

[a]Two-fold excess amounts of solvents and reagents were used.

Table 12. Examination of methoxy[11C]carbonylation reaction.

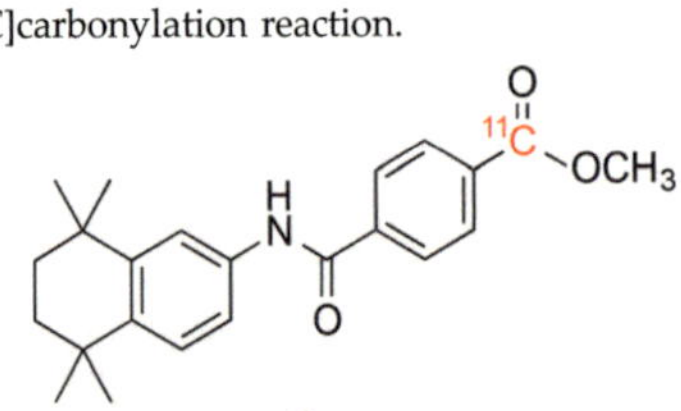

[11C]-**108**

proceeding at ambient temperature and under atmospheric pressure using conventional helium gas as a [11C]CO carrier, if successful, should be highly attractive to organic chemists who desire to synthesize [11C]carbonyl compounds such as 11C-labeled amides, ureas, isocyanates, oxazoles, and carbamates directly. Another approach to perform [11C]carbonylation of aryl halides and triflates at atmospheric pressure is reported by Dahl et al. [93], using a hindered phosphine ligand, xantphos, in combination with palladium(π-cinnamyl) chloride dimer. Recently, Andersen et al. reported that the isolation of xantphos ligated palladium-aryl oxidative addition complexes, whose generation generally seems the rate determining step, and [11C]carbonylation

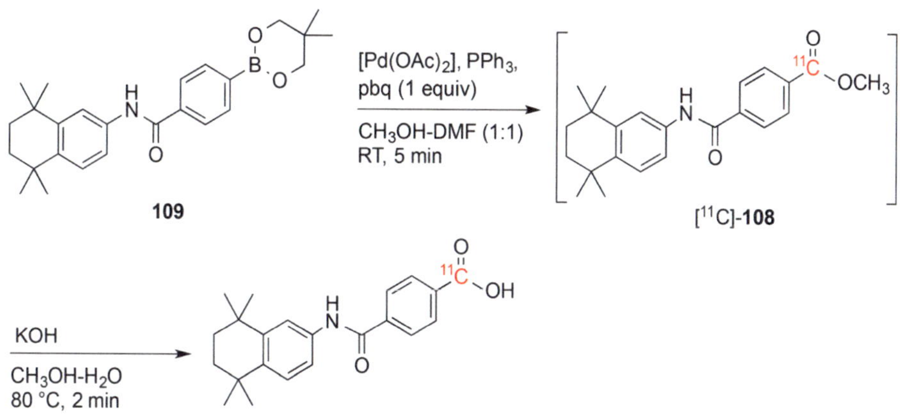

[11C]-110

Scheme 17. One-pot synthesis of [^{11}C]Am80 ([^{11}C]-**110**).

[^{11}C]aspirin ([^{11}C]-**111**) [^{11}C]salicylamide ([^{11}C]-**112**) [^{11}C]nicotinamide ([^{11}C]-**113**)

with [^{11}C]CO under low pressure have been successfully accomplished to give [^{11}C]raclopride and [^{11}C]olaparib with excellent radiochemical purity and yield [93]. Rahman et al. reported nickel-mediated [^{11}C]aminocarbonylation of non-activated alkyl iodides containing β-hydrogen using low concentrations of [^{11}C]CO in the nonpolar protic solvent *tert*-butyl alcohol [94]. See also recent reviews for rapid C-[^{11}C]carbonylation [95].

3.3.1. *Rapid C-[^{11}C]carboxylation under ambient temperature and atmospheric pressure*

Transition-metal-mediated rapid [^{11}C]carboxylation using [^{11}C]CO_2 and organoboronate as a stable metalloid substrate is favorable for ^{11}C-labeling of carboxyl groups. Pike et al. reported Cu^{I}-mediated rapid [^{11}C]carboxylation in the presence of KF/kryptofix 2.2.2 [96]. Alternatively, we found that the use of acetone is more effective for [^{11}C]CO_2 trapping at temperature ranging

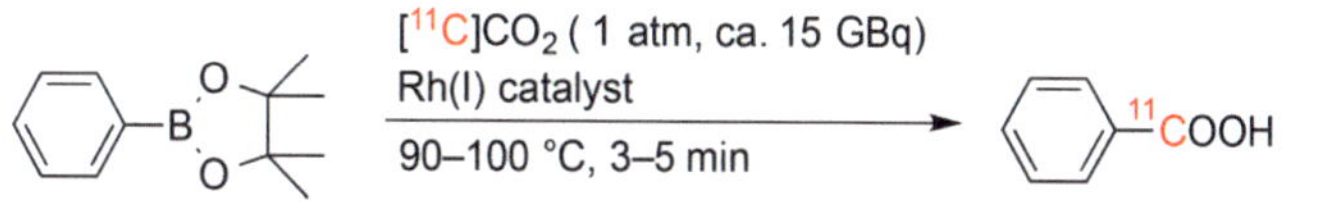

75

[^{11}C]-**114**
Isolated radioactivity: 2–3 GBq
[^{11}C]CO_2-based DCY: 38–57%

Scheme 18. Rh^{I}-mediated rapid [^{11}C]carboxylation using [^{11}C]CO_2 and organoboronate.

from −78 to −98°C than the use of molecular sieves as absorbent materials. Consequently, [^{11}C] carboxylation was promoted at atmospheric pressure with a short reaction time of 3–5 min using Rh^{I} catalyst (**Scheme 18**). The desired [^{11}C]benzoic acid ([^{11}C]-**114**) was obtained with moderate radioactivity (2–3 GBq) at 38–57% DCY using approximately 15 GBq of [^{11}C]CO_2 [97].

3.4. Other opportunities for rapid ^{11}C-labeling: Synthesis of [^{11}C]NSAIDs and their esters by rapid *C*-methylation of enolates

In order to perform the *in vivo* molecular imaging of cyclooxygenases (COXs), well-known as the key enzymes in prostaglandin biosynthesis, we intended to develop a novel method to rapidly incorporate a ^{11}C radionuclide into various 2-arenylpropionic acids that have a common methylated structure, particularly abundant among nonsteroidal anti-inflammatory drugs (NSAIDs). Consequently, we elaborated the rapid ^{11}C-labeling using the reaction of [^{11}C]CH_3I and an enolate intermediate generated from the corresponding ester under basic conditions, followed by one-pot hydrolysis to convert it into the ^{11}C-incorporated acid as [^{11}C]NSAID (**Figure 26A**) [98]. Methoxy 2-arenylpropionate (**115**) is much less polar due to the increase in hydrophobicity of an introduced methyl group and the less hyperconjugation between the C—H σ bond of the benzylic position and C=O π*, which is also possible for the LUMO (π*) of a phenyl moiety, allowing easy separation of the desired ^{11}C-labeled product from the demethylated compound. This method is quite general and utilized for the syntheses of the following six PET probes of NSAIDs: [^{11}C]ibuprofen ([^{11}C]-**122**), [^{11}C]naproxen ([^{11}C]-**123**), [^{11}C]flurbiprofen ([^{11}C]-**124**), [^{11}C]fenoprofen ([^{11}C]-**125**), [^{11}C] ketoprofen ([^{11}C]-**126**), [^{11}C]loxoprofen ([^{11}C]-**127**), and their corresponding esters as racemates ([^{11}C]-**116–121**), with sufficient radioactivity (1.7–5.5 GBq) for animal and human PET studies. The isolated radiochemical yields (decay-corrected) based on [^{11}C]CH_3I of [^{11}C]-**116–127** were 26–76%. Notably, we found that methyl esters were particularly useful as pro-radioprobes for the study of neuroinflammation in the brain. The microPET studies of rats with lipopolysaccharide (LPS)-induced brain inflammation clearly showed that the radioactivity of the PET probes, [^{11}C] ketoprofen methyl ester ([^{11}C]-**120**), and [^{11}C]ketoprofen ([^{11}C]-**126**) specifically accumulated in the inflamed region (**Figure 26B**), giving the first successful example of the *in vivo* molecular imaging of neuroinflammation by noninvasive PET technology. A metabolite analysis of the rat brain revealed that the intravenously administrated methyl ester was initially taken up in the brain and then underwent hydrolysis to form a pharmacologically active form of the corresponding acids. Hence, we succeeded in the general ^{11}C-labeling of 2-arenylpropionic acids and their methyl esters as PET probes of NSAIDs to construct a potentially useful PET-probe library for the *in vivo* imaging of inflammation involved in the COX expression [99]. The aforementioned racemic NSAD methyl ester is readily separated by a chiral column to give optically pure compounds among which *S*-enantiomer of [^{11}C]ketoprofen methyl ester ((*S*)-[^{11}C]-**120**) was found to be a highly selective PET probe for COX-1. Studies using APP-Tg mice demonstrated that (*S*)-[^{11}C]-**126** detected COX-1 in activated microglia that are associated with amyloid plague progression, suggesting the involvement of COX-1 in the neuro-inflammation process of Alzheimer's disease [99]. [^{11}C]Ketoprofen methyl ester has been used for human whole body bio-distribution for imaging of neuroinflammation [100]. Tetrabutylammonium fluoride (TBAF) was also effective to promote the rapid [^{11}C]methylation of the enolate in THF as found in our and relative groups [101].

http://dx.doi.org/10.5772/intechopen.72868

(A)

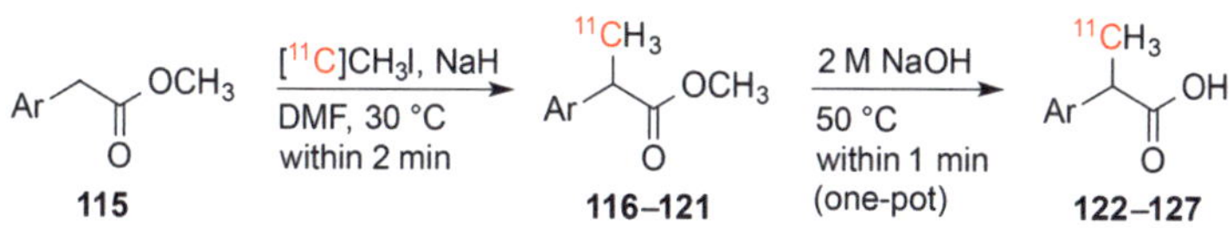

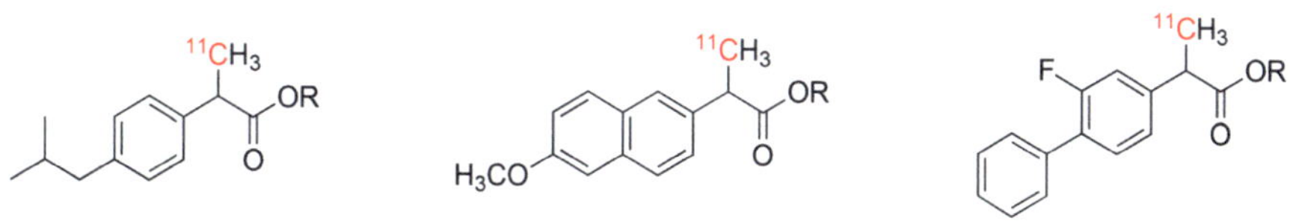

[^{11}C]**116**: R = CH_3, [^{11}C]ibuprofen methyl ester	[^{11}C]**117**: R = CH_3, [^{11}C]naproxen methyl ester	[^{11}C]**118**: R = CH_3, [^{11}C]flurbiprofen methyl ester
[^{11}C]**122**: R = H, [^{11}C]ibuprofen	[^{11}C]**123**: R = H, [^{11}C]naproxen	[^{11}C]**124**: R = H, [^{11}C]flurbiprofen

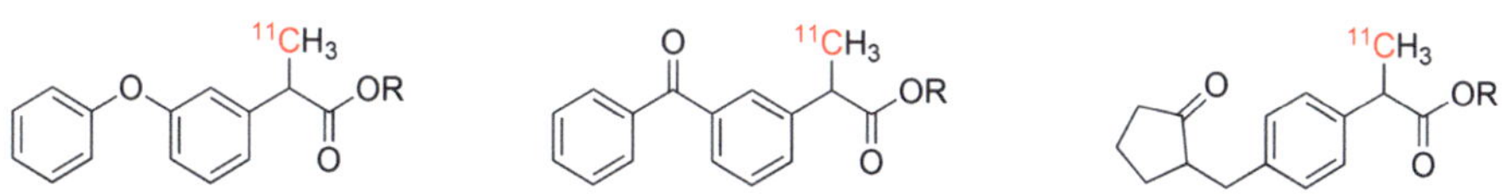

[^{11}C]**119**: R = CH_3, [^{11}C]fenoprofen methyl ester	[^{11}C]**120**: R = CH_3, [^{11}C]ketoprofen methyl ester	[^{11}C]**121**: R = CH_3, [^{11}C]loxoprofen methyl ester
[^{11}C]**125**: R = H, [^{11}C]fenoprofen	[^{11}C]**126**: R = H, [^{11}C]ketoprofen	[^{11}C]**127**: R = H, [^{11}C]loxoprofen

(B)

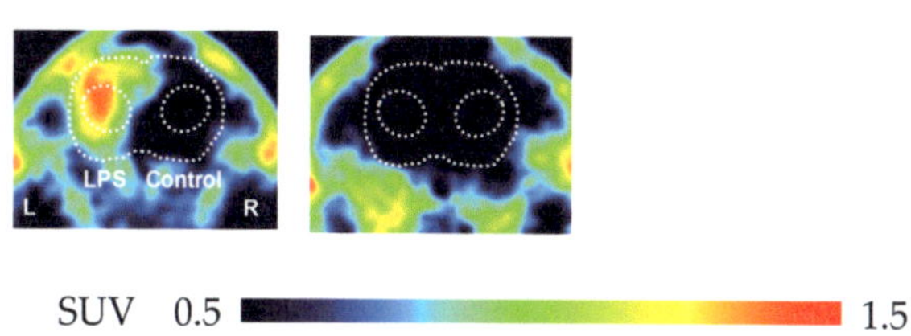

Figure 26. Syntheses of ^{11}C-labeled 2-arylpropionoc acids and their esters (A), and PET images of [^{11}C]ketoprofen methyl ester ([^{11}C]-**120**, left panel) and [^{11}C]ketoprofen ([^{11}C]-**126**, right panel) in rat brain inflammation induced by lipopolysaccharide injection into the left striatum (B). Left PET image shows high accumulation in the area of inflammation, indicating that the methyl ester penetrated the blood–brain barrier and underwent hydrolysis in the brain to produce carboxylic acid as a pharmacologically active form, eventually accumulating in the inflammation area.

4. Conclusions

PG synthesis: Organometallic techniques opened a direct way to PGs. The one-pot three-component coupling comprising the construction of the whole PG C_{20}-framework by the combination of the organocopper or organozincate conjugate addition of the ω-side chain to (4*R*)-4-hydroxy-2-cyclopentenone and selective monoalkylation with the α-side chain unit was realized by the addition of 1 equiv of $(C_6H_5)_3SnCl$ or $(CH_3)_2Zn$ and excess HMPA for the first time by our group. The use of acetylenic halide as an α-side chain led to the 5,6-didehydro-PGE_2 derivative, which served as the common intermediate for general synthesis of natural PGs. Combination of a controlled radical reaction, photochemical process, and organometallic and organosilicon chemistry allowed the synthesis of isocarbacyclin, a stable prostacyclin analog. 15*S*-APNIC with an azidophenyl group as a photoreactive function was also synthesized; it has a high affinity to the prostacyclin receptor protein in mastocytoma P-815 cells (IC_{50} = 3 nM). A protein of approximately 45 kDa was labeled by photolysis using (15*S*)-[^{3}H]APNIC incubated with membranes of porcine platelets including a PGI_2 receptor. Likely, 15*R*-TIC with a tolyl group in ω-side chain was synthesized from the same aldehyde intermediate, which selectively bound to a novel PGI_2 receptor subtype, IP_2, expressed in the CNS (IC_{50} = 32 nM). The length of the ω-side chain and the position of the methyl substituent on the aromatic ring strongly influenced the binding characteristics. 15-deoxy-TIC lacking a hydroxy group at C(15) showed ten-fold higher affinity for the IP_2. The green synthesis of the key intermediate, 5,6-didehydro-PGE_2 derivative for general synthesis of natural PGs described was achieved ultimately by efficient one-pot three-component coupling consisting of organozincate conjugate addition followed by the enolate trapping with an organic triflate using stoichiometric amounts of three-components (enone and two ω- and α-side chains with nearly 1:1:1 ratio) without HMPA and heavy metals. The intermediates corresponding to the syntheses of 15*R*–TIC and 15-deoxy-TIC, respectively, were also synthesized by similar protocols.

PET probe synthesis: A groundbreaking methodology based on the use of advanced chemistry was introduced for the synthesis of short-lived ^{11}C-incorporated PET probes. First, a general method for the rapid cross-coupling reaction of methyl iodide with phenyltributylstannane (Stille-type rapid cross-coupling reaction) was established, producing a methylated product at high yield in the presence of a bulky tri-*o*-tolylphosphine-bound coordinatively unsaturated Pd^0 complex, a Cu^I salt, and K_2CO_3 in DMF. The two salts work synergically to promote the Sn/Cu transmetallation. The reaction was efficiently used for the synthesis of the 15*R*-[^{11}C]TIC methyl ester as a PET probe. The rapid *C*-methylation was expanded to other types of rapid *C*-methylations including the methylations of heteroaromatic compounds and alkenes by using other Cu^I/CsF synergy as well as alkynes and choosing a bulky trialkylphosphine/KF or CsF in the absence of Cu^I salt. The choice of NMP, as a solvent, is important in the methylation of heteoaromatic compounds. These rapid reactions could afford various radiolabeled biologically and clinically important molecules. To meet further demands for efficient labeling methods in PET, a rapid *C*-methylation was established using methyl iodide and organoboranes (Suzuki-Miyaura type rapid cross-coupling reaction), complementary to organostannanes. The reactions using phenyl, heteroaromatic, and alkenylborane substrates proceeded at high yields in the presence of $[Pd_2(dba)_3]$, tri-*o*-tolylphosphine, and K_2CO_3 or K_2CO_3/H_2O, while the rapid cross-coupling reactions of sp^3- and

http://dx.doi.org/10.5772/intechopen.72868

Other novel PET probes synthesized by us and our collaborators:

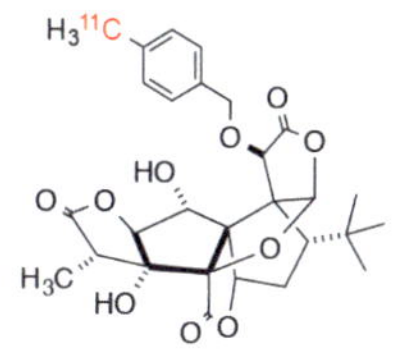

[^{11}C]methylbenzyl GB ([^{11}C]-**128**)

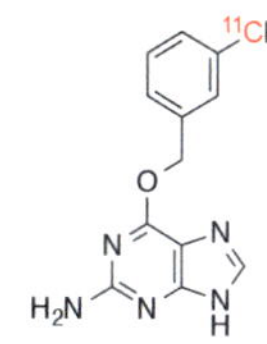

[^{11}C]methylbenzyl guanine ([^{11}C]-**129**)

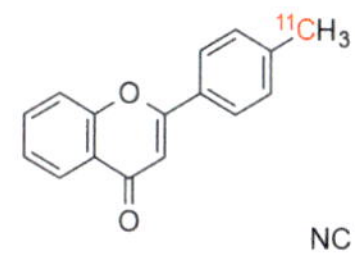

4'-[^{11}C]methylflavone ([^{11}C]-**130**)

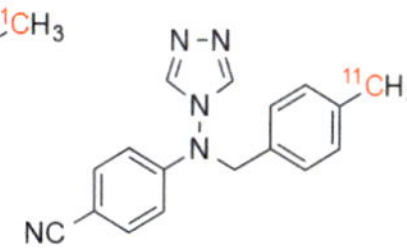

[^{11}C]cetrozole ([^{11}C]-**131**)

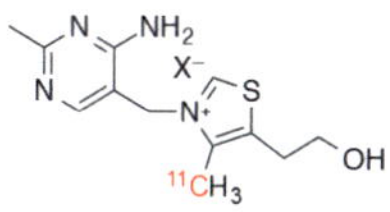

[^{11}C]thiamine ([^{11}C]-**132**)
X is expected to be OH⁻ after ion exchange.

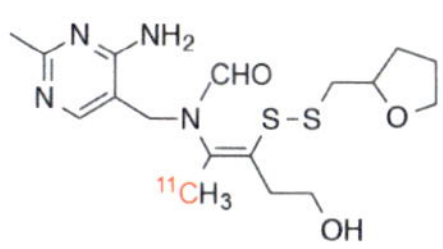

[^{11}C]fursultiamine ([^{11}C]-**133**)

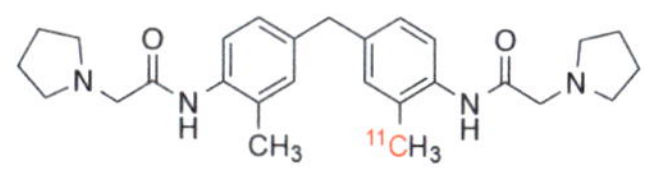

^{11}C-labeled GN8 ([^{11}C]-**134**)

Various PET probes synthesized by other groups by the use of our original conditions or modified conditions:

[^{11}C]MADAM ([^{11}C]-**135**)

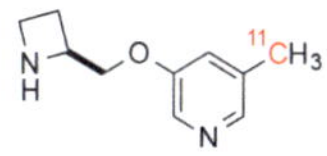

[^{11}C]5MA ([^{11}C]-**136**)

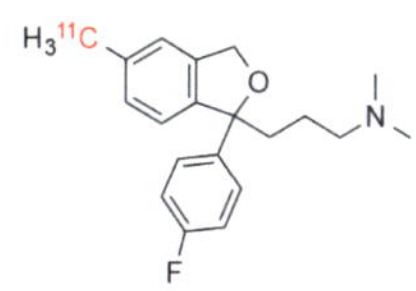

[5-methyl-^{11}C]{3-[1-(4-fluorophenyl)-5-methyl-1,3-dihydroisobenzofuran-1-yl]-propyl}-dimethylamine ([^{11}C]-**137**)

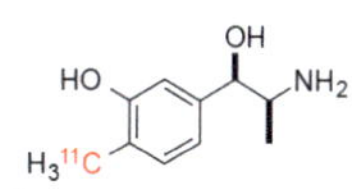

4-[^{11}C]methylmetaraminol ([^{11}C]-**138**)

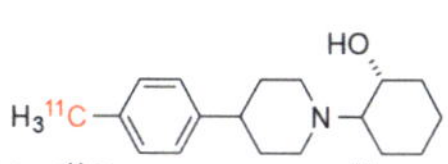

(+)-*p*-[^{11}C]methylvesamicol ([^{11}C]-**139**)

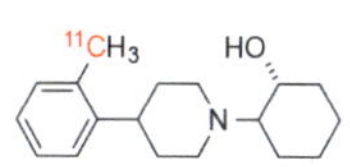

(−)-[^{11}C]OMV ([^{11}C]-**140**)

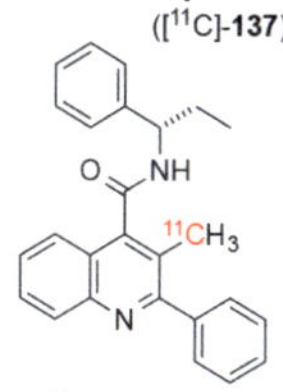

[^{11}C]SB 222200 ([^{11}C]-**141**)

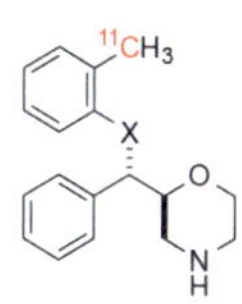

[^{11}C]MENET (X = O) ([^{11}C]-**142a**)
[^{11}C]MESNET (X = S) ([^{11}C]-**142b**)

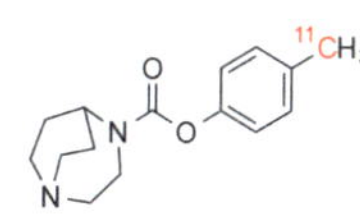

[^{11}C]CHIBA-1001 ([^{11}C]-**143**)

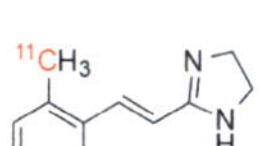

[^{11}C]metrazoline ([^{11}C]-**144**)

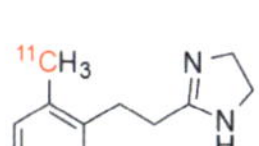

[^{11}C]TEIMD ([^{11}C]-**145**)

[^{11}C]FTIMD ([^{11}C]-**146**)

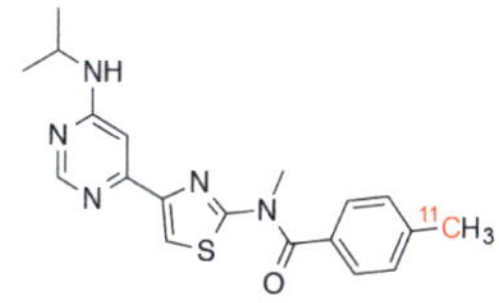

N-[4-[6-(isopropylamino)pyrimidin-4-yl]-1,3-thizol-2-yl]-*N*-methyl-4-[^{11}C]methylbenzamide ([^{11}C]-**147**)

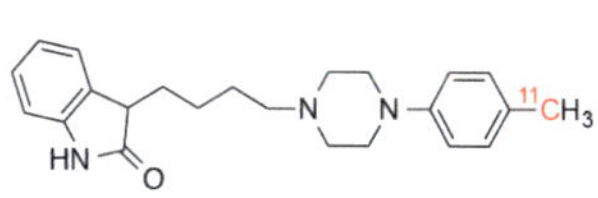

[^{11}C]CIMBI-712 ([^{11}C]-**148**)

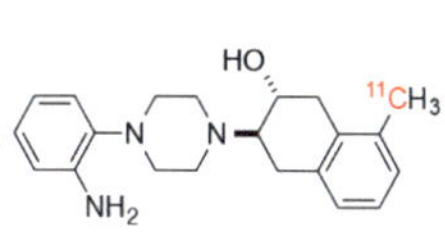

[^{11}C](*R*,*R*)HAPT ([^{11}C]-**149**)

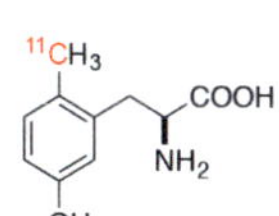

[^{11}C]6Me*m*Tyr ([^{11}C]-**150**)

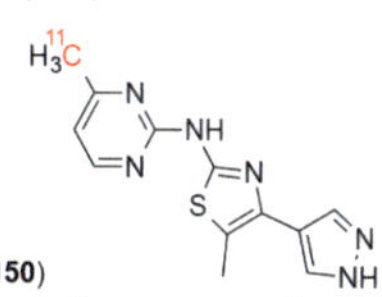

[^{11}C]ADX88178 ([^{11}C]-**151**)

sp^3-hybridized carbons were possible using bulky tri-*tert*-butylphosphine-bound coordinatively unsaturated Pd^0 complex and CsF in 90:10 DMF/H_2O (*v*/*v*) at high yields. The incorporation of rather longer half-life ^{18}F radionuclides into organic structures was also possible by rapid *C*-[^{18}F] fluoromethylation using [^{18}F]FCH_2X (X = I or Br) and organoboranes. It was also found that the conditions reported so far by other researchers using the Stille and Suzuki-Miyaura type cross-coupling reactions using organostannane and borane compounds (even using TlOH as base) were totally much less effective than ours. The rapid *C*-[^{11}C]methylation conditions would be useful for [^{11}C]CH_3I prepared by gas-phase reaction of [^{11}C]CH_4 and iodine [102, 103], providing the labeled compound with higher specific activity. Rapid *C*-[^{11}C]carbonylation was also established by the Pd^{II}-mediated reaction of [^{11}C]CO and common arylboronate at ambient temperature under atmospheric pressure using conventional helium carrier gas. Likely, the rapid *C*-[^{11}C]carboxylation under atmospheric pressure was promoted by the use of Rh^I catalyst.

Recently, we have further applied our rapid *C*-methylation to the synthesis of 10-*O*-*p*-[^{11}C] methylbenzyl ginkgolide B (**128**) and O^6-[(3-[^{11}C]methyl)benzyl]guanine (**129**) as newly designed PET probes for the study of brain permeability of ginkgolides in monkeys and rats [104] and for conducting the imaging of DNA repair protein O^6-methylguanine-DNA methyltransferase in glioblastoma [105], respectively. Further, novel rapid *C*-methylation methodology have also been employed for the syntheses of novel PET probes by us and our collaborators, such as 4′-[^{11}C] methylflavone ([^{11}C]-**130**) [106], [^{11}C]cetrozole (^{11}C-incorporated aromatase inhibitor) ([^{11}C]-**131**) [107], [^{11}C]thiamine ([^{11}C]-**132**) [108], [^{11}C]fursultiamine ([^{11}C]-**133**) [108], and ^{11}C-labeled GN8 ([^{11}C]-**134**) for prion imaging [109]. Furthermore, our new methods have widely been applied by other research groups to synthesize ^{11}C-labeled PET probes such as a serotonin transporter inhibitor, [^{11}C]MADAM ([^{11}C]-**135**) [110], [^{11}C]toluene ([^{11}C]-**46**) [111]; a central nicotinic acetylcholine inhibitor, [^{11}C]5MA ([^{11}C]-**136**) [112]; a serotonin reuptake inhibitor, citalopram analog, [5-methyl-^{11}C]{3-[1-(4-fluorophenyl)-5-methyl-1,3-dihydroisobenzofuran-1-yl]- propyl}-dimethylamine ([^{11}C]-**137**) [43]; an adrenergic neurotransmitter, 4-[^{11}C]methylmetaraminol ([^{11}C]-**138**) [113]; a COX-2 selective inhibitor and prescription drug, [^{11}C]celecoxib ([^{11}C]-**82**) [46]; $sigma_1$ receptor ligand, (+)-*p*-[^{11}C]methylvesamicol ([^{11}C]-**139**) [114]; vesicular acetylcholine transporter ligand, (−)-[^{11}C]OMV ([^{11}C]-**140**) [115]; an NK-3 receptor antagonist, [^{11}C]SB 222200 ([^{11}C]-**141**) [116]; a norepinephrine transporter ligand, a reboxetin analog, [^{11}C]MENET ([^{11}C]-**142a**) and [^{11}C]MESNET ([^{11}C]-**142b**) [117]; a derivative of the selective $\alpha7$ nicotinic acetylcholine receptor partial agonist, [^{11}C]CHIBA-1001 ([^{11}C]-**143**) [118]; an I_2-imidazoline receptor selective ligand, [^{11}C]metrazoline ([^{11}C]-**144**) [119], [^{11}C]TEIMD ([^{11}C]-**145**) [119], and [^{11}C]FTIMD ([^{11}C]-**146**) [119]; a metabotropic glutamate 1 receptor ligand, *N*-[4-[6-(isopropylamino)pyrimidin-4-yl]-1,3-thizol-2-yl]-*N*-methyl-4-[^{11}C]methylbenzamide ([^{11}C]-**147**) [120]; a selective 5-HT_7 antagonist, [^{11}C]CIMBI-712 ([^{11}C]-**148**) [121]; vesicular acetylcholine transporter ligand, [^{11}C](*R*,*R*)HAPT ([^{11}C]-**149**) [122]; imaging of presynaptic dopamine synthesis, [^{11}C]6Me*m*Tyr ([^{11}C]-**150**) [123]; imaging of metabotropic glutamate receptor subtype 4, [^{11}C]ADX88178 ([^{11}C]-**151**) [124], etc.

5. Perspectives

The straightforward construction of the main skeleton of a molecule by assembling multi-components (convergent synthesis) is an ideal synthetic methodology in organic synthesis in

http://dx.doi.org/10.5772/intechopen.72868

view of its directness and high synthetic flexibility. Such a synthetic approach requires readily accessible optically active components and highly selective asymmetric or diastereomeric C—C bond forming reactions. Three-component PG synthesis established in this study meets both requirements [15]. It is possible to expand the short-step synthesis method with high flexibility to the synthesis of a variety of analogs with improved biological activities and new significant biofunctions. In fact, it was well demonstrated that stable PG analogs thus created served not only to elucidate the target molecule connected with the mechanism of biofunction but also to discover new targets with novel biofunctions such as the IP_2 receptor in the brain. The green process has a great potential from practical and environmental points of view, and therefore, this convergent process would also be applied to a wide variety of enone structures to supply important organic frameworks for the syntheses of molecules with biological significance.

PET molecular imaging is the only non-invasive method for elucidating the entire-body pharmacokinetics of molecules in humans with high sensitivity and accurate quantification. This technique could be adaptable to the screening of drug candidates by introducing human microdosing studies at an early stage of the drug development (phase 0 by microdose administration under regulated guidance) [125] to decrease the large drop out of drug candidates (>90%, called "the Death Valley") during clinical trials (phases I–III), eventually saving huge investment and time for drug development (**Figure 27**); thus, revolutionizing drug development process and disease diagnosis. Furthermore, the use of a defined PET biomarker would be effective for the evaluation of drug efficacy after long-termed administration of a drug candidate particularly for dementia patients such as those suffering from Alzheimer's disease. Thus, novel chemical probes with a short-lived radioisotope would serve as efficient tools to particularly advance a study in human and environmental sciences in terms of ultra high sensitivity and safety due to less radiation exposure.

Our new methods are also applicable to other C, H, and F isotopes, such as $^{13}CH_3$, $^{14}CH_3$, CD_3, and $CH_2{}^{19}F$, allowing the synthesis of a molecular probe for accelerator mass spectrometry (AMS), MRI, etc.

An interdisciplinary associated diverse study among chemistry, biology, and medicine has increasingly become more important in the field of chemical biology and molecular imaging particularly relating to drug development and disease diagnosis.

Chemical tools covering a broad range of design, synthesis, and the labeling of significant biological and pharmaceutical molecules could play a central role in an interdisciplinary research area such as bio-, medical-, life (live)-, and environmental sciences.

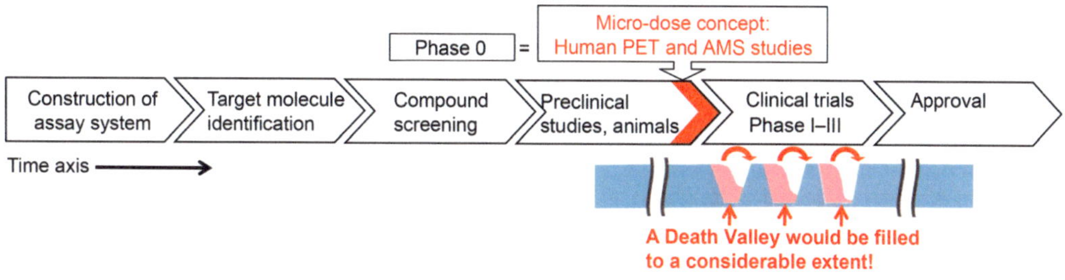

Figure 27. The use of PET toward revolutionizing drug discovery processes.

Acknowledgements

We would like to express our sincere gratitude to Prof. Ryoji Noyori (Nagoya University, RIKEN, and now Director-General of Center for Research and Development Strategy, JST) for the promotion of PG synthesis and PET molecular imaging science. We also thank Drs. Yanagisawa A, Morita Y, Koyano H as well as Profs. Seiji Ito (Kansai Medical University) and Manabu Negishi, and Atsushi Ichikawa (Kyoto University) and their colleagues for the studies of the synthesis of prostaglandins (PGs) and the evaluation of their biological activities. We are also grateful to Prof. Bengt Långström (Uppsala University) for the PET study of TIC as an international collaboration between Sweden and Japan. Drs. Hosoya T, Wakao M, Sumi K, Siqin, Takashima-Hirano M, Furuta K, Kanazawa M, Ijuin R, Zhang Z, and other colleagues, whose names are listed in the references, have contributed as chemists in rapid *C*-[^{11}C]methylation, *C*-[^{18}F]fluoromethylation, and *C*-[^{11}C]carbonylation projects and applications for the synthesis of various PET probes. We also thank Prof. Yasuyoshi Watanabe and Dr. Hirotaka Onoe (RIKEN Center for Life Science Technologies (CLST)) as director and leader, respectively, and their colleagues listed in original papers on PG biological studies, particularly, 15*R*-TIC and the promotion of PET imaging studies. We also thank Prof. Yuichi Sugiyama (RIKEN Innovation Center (RlnC)) for his active guidance in evaluating the drug behavior of transporters in organs. Prof. Koji Nakanishi (Columbia University) and Dr. Hideo Tsukada (Hamamatsu Photonics K. K.) for the study on the brain permeability of ^{11}C-labeled ginkgolide and Profs. Toshihiko Wakabayashi and Atsushi Natsume (Nagoya University) and Dr. Kengo Ito (National Center for Geriatrics and Gerontology, Japan) for the synthesis of ^{11}C-labeled guanine derivative as a PET probe for MGMT imaging in glioblastoma are also appreciated. These studies were supported in part by a Grant-in-Aid for Creative Scientific Research (No. 13NP0401), Scientific Research (A) (Nos. 24249073 and 25242070), a consignment expense for the Molecular Imaging Program on "Research Base for Exploring New Drugs" (2005–2009), Japan Advanced Molecular Imaging Program (2010–) from the Ministry of Education, Culture, Sports, Science and Technology (MEXT) of Japan, and Basic Science and Platform Technology Program for Innovative Biological Medicine (JP17am0301022) from Japan Agency for Medical Research and Development (2015–).

Author details

Masaaki Suzuki[1]*, Hiroko Koyama[2], Hideki Ishii[3], Koichi Kato[4], Margareta Ögren[5] and Hisashi Doi[6]

*Address all correspondence to: suzukims@ncgg.go.jp

1 National Center for Geriatrics and Gerontology, Japan

2 Gifu University, Japan

3 National Institutes for Quantum and Radiological Science and Technology, Japan

4 National Center of Neurology and Psychiatry, Japan

5 Institution of Radiation Science, Umeå University Hospital, Sweden

6 RIKEN Center for Life Science Technologies, Japan

References

[1] Suzuki M, Noyori R, Långström B, Watanabe Y. Molecular design of prostaglandin probes in brain research: High, specific binding to a novel prostacyclin receptor in the central nervous system. Bulletin of the Chemical Society of Japan. 2000;**73**:1053-1070. DOI: 10.1246/bcsj.73.1053

[2] Suzuki M, Doi H, Hosoya T, Långström B, Watanabe Y. Rapid methylation on carbon frameworks leading to the synthesis of a PET tracer capable of imaging a novel CNS-type prostacyclin receptor in living human brain. Trends in Analytical Chemistry. 2004;**23**:595-607. DOI: 10.1016/j.trac.2004.06.003

[3] Suzuki M, Doi H. Pd^0-mediated rapid *C*-[^{11}C]methylations and *C*-[^{18}F]fluoromethylations: Revolutionary new methodologies for the synthesis of short-lived PET molecular probes. Journal of Synthetic Organic Chemistry, Japan. 2010;**68**:1195-1206. DOI: 10.5059/yukigoseikyokaishi.68.1195

[4] Suzuki M, Koyama H, Takashima-Hirano M, Doi H. Pd^0-mediated rapid *C*-[^{11}C]methylation and *C*-[^{18}F]fluoromethylation: Revolutionary advanced methods for general incorporation of short-lived positron-emitting ^{11}C and ^{18}F radionuclides in an organic framework. In: Hsieh C-H, editor. Positron Emission Tomography-Current Clinical and Research Aspects. Rijeka, Croatia: InTech; 2012. pp. 115-152. DOI: 10.5772/31778. Chap. 5

[5] Suzuki M, Doi H, Koyama H, Zhouen Z, Hosoya T, Onoe H, Watanabe Y. Pd^0-mediated rapid cross-coupling reactions, the rapid *C*-[^{11}C]methylations, revolutionarily advancing the syntheses of short-lived PET molecular probes. The Chemical Record. 2014;**14**: 516-541. DOI: 10.1002/tcr.201400002

[6] Noyori R, Suzuki M. Organic synthesis of the prostaglandins. Science. 1993;**259**:44-45. DOI: 10.1126/science.8418493, and references cited therein; see also Advances in Prosta-Glandins, Thromboxane, and Leukotriene Research. Vols. 11–21. New York: Raven; 1983–1991

[7] Corey EJ. Studies on the total synthesis of prostaglandins. Angewandte Chemie International Edition in English. 1991;**30**:455-465. DOI: 10.1002/anie.199104553

[8] Noyori R, Suzuki M. Prostaglandin synthesis by three-component coupling. Angewandte Chemie International Edition in English. 1984;**23**:847-876. DOI: 10.1002/anie.198408471

[9] Suzuki M, Yanagisawa A, Noyori R. An extremely short way to prostaglandins. Journal of the American Chemical Society. 1985;**107**:3348-3349. DOI: 10.1021/ja00297a051

[10] Morita Y, Suzuki M, Noyori R. An organozinc aid in alkylation and acylation of lithium enolates. The Journal of Organic Chemistry. 1989;**54**:1785-1787. DOI: 10.1021/jo00269a006

[11] Suzuki M, Koyama H, Noyori R. Effects of HMPA on the structure and reactivity of the lithium enolate of cyclopentanone: The dimer is responsible for alkylation and proton exchange reactions. Bulletin of the Chemical Society of Japan. 2004;**77**:259- 268. DOI: 10.1246/bcsj.77.259

[12] Suzuki M, Suzuki T, Kawagishi T, Morita Y, Noyori R. Conjugate addition of phosphine-complexed organocopper reagents to α,β-unsaturated ketones. Israel Journal of Chemistry. 1984;**24**:118-124. DOI: 10.1002/ijch.198400021

[13] Suzuki M, Yanagisawa A, Noyori R. The three-component coupling synthesis of prostaglandins. Journal of the American Chemical Society. 1988;**110**:4718-4726. DOI: 10.1021/ja00222a033

[14] Suzuki M, Morita Y, Koyano H, Koga M, Noyori R. Three-component coupling synthesis of prostaglrndins. a simplified general procedure. Tetrahedron. 1990;**46**:4809-4822. DOI: 10.1016/S0040-4020(01)85596-4

[15] Noyori R, Suzuki M. An organometallic way to prostaglandins: The three-component coupling synthesis. Chemtracts. Organic Chemistry. 1990;**3**:173-197

[16] Suzuki M, Kawagishi T, Yanagisawa A, Suzuki T, Okamura N, Noyori R. Three-component coupling synthesis of prostaglandins: The aldol route. Bulletin of the Chemical Society of Japan. 1988;**61**:1299-1312. DOI: 10.1246/bcsj.61.1299

[17] Tanaka T, Hazato A, Bannai K, Okamura N, Sugiura S, Manabe K, Kurozumi S, Suzuki M, Noyori R. Short synthesis of 6-oxoprostaglandin E_1 and 6-oxoprostaglandin $F_1\alpha$. Tetrahedron Letters. 1984;**25**:4947-4950. DOI: 10.1016/S0040-4039(01)91266-3

[18] Suzuki M, Mori M, Niwa T, Hirata R, Furuta K, Ishikawa T, Noyori R. Chemical implications for antitumor and antiviral prostaglandins: Reaction of Δ^7-prostaglandin A_1 and prostaglandin A_1 methyl esters with thiols. Journal of the American Chemical Society. 1997;**119**:2376-2385. DOI: 10.1021/ja9628359; Furuta K, Tomokiyo K, Satoh T, Watanabe Y, Suzuki M. Designed prostaglandins with neurotrophic activities. ChemBioChem. 2000;**1**:283-286. DOI: 10.1002/1439-7633(20001117)1:4<283::AID-CBIC283>3.0.CO;2-O; Satoh T, Furuta K, Tomokiyo K, Namura S, Nakatsuka D, Sugie Y, Ishikawa Y, Hatanaka H, Suzuki M, Watanabe Y. Neurotrophic actions of novel compounds designed from cyclopentenone prostaglandins. Journal of Neurochemistry. 2001;**77**:50-62. DOI: 10.1046/j.1471-4159.2001.00229.x

[19] Suzuki M, Kiho T, Tomokiyo K, Furuta K, Fukusima S, Takeuchi Y, Nakanishi M, Noyori R. Rational design of antitumor prostaglandins with high biological stability. Journal of Medicinal Chemistry. 1998;**41**:3084-3090. DOI: 10.1021/jm9801657

[20] Shibasaki M, Torisawa Y, Ikegami S. Synthesis of 9(0)-methano-$\Delta^{6(9\alpha)}$-PGI_1: The highly potent carbon analog of prostacyclin. Tetrahedron Letter. 1983;**24**:3493-3496. DOI: 10.1016/S0040-4039(00)86021-9

[21] Suzuki M, Koyano H, Noyori R. A controlled synthesis of isocarbacyclin. The Journal of Organic Chemistry.1987;**52**:5583-5588. DOI: 10.1021/jo00234a014

[22] Suzuki M, Koyano H, Noyori R, Hashimoto H, Negishi M, Ichikawa A, Ito S. An azido-functionalized isocarbacyclin analogue acting as an efficient photoaffinity probe for a prostacyclin receptor. Tetrahedron. 1992;**48**:2635-2658. DOI: 10.1016/S0040-4020(01)88526-4

[23] Sodeoka M, Shibasaki M. Practical synthesis of (+)-9-(0)-methano-$\Delta^{6(9\alpha)}$-PGI_1. Chemistry Letters. 1984;579-582. DOI: 10.1246/cl.1984.579; Sodeoka M, Ogawa Y, Mase T, Shibasaki M. Efficient synthesis of isocarbacyclins. Chemical and Pharmaceutical Bulletin.1989;**37**:586-598. DOI: 10.1248/cpb.37.586; Sodeoka M, Ogawa Y, Kirio Y, Shibasaki M. Stereocontrolled synthesis of exocyclic olefins using arene tricarbonyl chromium complex-catalyzed hydrogenation. I. Efficient synthesis of carbacyclin and its analogs. Chemical and Pharmaceutical Bulletin. 1991;**39**:309-322. DOI: 10.1248/cpb.39.309

[24] Ito S, Hashimoto H, Negishi M, Suzuki M, Koyano H, Noyori R, Ichikawa A. Identification of the prostacyclin receptor by use of [15-$^3H_1$]-19-(3-azidophenyl)-20-norisocarbacyclin, an irreversible specific photoaffinity probe. The Journal of Biological Chemistry. 1992;**267**:20326-20330

[25] Suzuki M, Kato K, Noyori R, Watanabe Yu, Takechi H, Matsumura K, Långström B, Watanabe Y. (15*R*)-16-*m*-tolyl-17,18,19,20-Tetranorisocarbacyclin: A stable ligand with high binding affinity and specificity for a prostacyclin receptor in the central nervous system. Angewandte Chemie International Edition. 1996;**35**:334-336.DOI: 10.1002/anie.199603341

[26] Takechi H, Matsumura K, Watanabe Y, Kato K, Noyori R, Suzuki M, Watanabe Y. A novel subtype of the prostacyclin receptor expressed in the central nervous system. The Journal of Biological Chemistry. 1996;**271**:5901-5906. DOI: 10.1074/jbc.271.10.5901

[27] Suzuki M, Kato K, Watanabe Y, Satoh T, Matsumura K, Watanabe Y, Noyori R. 15-Deoxy-16-(*m*-tolyl)-17,18,19,20-tetranorisocarbacyclin: A simple TIC derivative with potent anti-apoptotic activity for neuronal cells. Chemical Communications. 1999:307-308. DOI: 10.1039/A807613H

[28] Satoh T, Ishikawa Y, Kataoka Y, Cui Y, Yanase H, Kato K, Watanabe Y, Nakadate K, Matsumura K, Hatanaka H, Kataoka K, Noyori R, Suzuki M, Watanabe Y. CNS-specific prostacyclin ligands as neuronal survival-promoting factors in the brain. The Journal of Neuroscience. 1999;**11**:3115-3124. DOI: 10.1046/j.1460-9568.1999.00791.x

[29] Cui Y, Kataoka Y, Satoh T, Yamagata A, Shirakawa N, Watanabe Yu, Suzuki M, Yanase H, Kataoka K, Watanabe Y. Protective effect of prostagrandin I_2 analogs on ischemic delayed neuronal damage in gerbils. Biochemical and Biophysical Research Communications. 1999;**265**:301-304. DOI: 10.1074/jbc.271.10.5901

[30] Gooding OW. An expedient triply convergent synthesis of prostaglandins. The Journal of Organic Chemistry. 1990;**55**:4209-4211. DOI: 10.1021/jo00300a048

[31] Lipshutz BH, Wood MR. A practical, general three-component coupling approach to prostaglandin and non-prostaglandin-related skeletal. Journal of the American Chemical Society. 1994;**116**:11689-11702. DOI: 10.1021/ja00105a010

[32] Current results accomplished by Koyama H, Izumiseki A, and Suzuki M in our Suzuki's research group. The details of these achievements have been filed as patent application: Suzuki M and Koyama H, JP2017-150745 prior to be reported in a chemical journal.

[33] Phelps ME, editor. Molecular Imaging and Its Biological Applications. New York/Berlin/Heidelberg: Springer; 2004. DOI: 10.1007/978-0-387-22529-6

[34] Lappin G, Garner RC. Big physics, small doses: The use of AMS and PET in human microdosing of development drugs. Nature Reviews. Drug Discovery. 2003;**2**:233-240. DOI: 10.1038/nrd1037

[35] Bergström M, Grahnén A, Eur LB. Positron emission tomography microdosing: A new concept with application in tracer and early clinical drug development. Journal of Clinical Pharmacology. 2003;**59**:357-366. DOI: 10.1007/s00228-003-0643-x

[36] Schubiger PA, Lehmann L, Friebe M, editors. PET Chemistry: The Driving Force in Molecular Imaging. New York/Berlin/Heidelberg: Springer; 2007. DOI: 10.1007/978-3-540-49527-7

[37] Sundoro-Wu BM, Schmall B, Conti PS, Dahl JR, Drumm P, Jacobsen JK. Selective alkylation of pyrimidyldianions: Synthesis and purification of ^{11}C labeled thymidine for tumor visualization using positron emission tomography. The International Journal of Applied Radiation and Isotopes. 1984;**35**:705-708. DOI: 10.1016/0020- 708X(84)90075-9

[38] Stille JK. The palladium-catalyzed cross-coupling reactions of organotin reagents with organic electrophiles. Angewandte Chemie International Edition. 1986;**25**:508-524. DOI: 10.1002/anie.198605081

[39] Morita DK, Stille JK, Norton JR. Methyl/phenyl exchange between palladium and a phosphine ligand. Consequences for catalytic coupling reactions. Journal of the American Chemical Society. 1995;**117**:8576-8581. DOI: 10.1021/ja00138a012

[40] Suzuki M, Doi H, Björkman M, Andersson Y, Långström B, Watanabe Y, Noyori R. Rapid coupling of methyl iodide with aryltributylstannanes mediated by palladium(0) complexes: A general protocol for the synthesis of $^{11}CH_3$-labeled PET tracers. Chemistry - A European Journal. 1997;**3**:2039-2042. DOI: 10.1002/chem.19970031219

[41] Paul F, Patt J, Hartwig JF. Structural characterization and simple synthesis of {Pd[P(*o*-Tol)$_3$]$_2$}. Spectroscopic study and structural characterization of the dimeric palladium(II) complexes obtained by oxidative addition of aryl bromides and their reactivity with amines. Organometallics. 1995;**14**:3030-3039. DOI: 10.1021/om00006a053

[42] Tolman CA. Steric effects of phosphorus ligands in organometallic chemistry and homogeneous catalysis. Chemical Reviews. 1977;**77**:313-348. DOI: 10.1021/cr60307a002

[43] Madsen J, Merachtsaki P, Davoodpour P, Bergström M, Långström B, Andersen K, Thomsen C, Martiny L, Knudsen GM. Synthesis and biological evaluation of novel carbon-11-labelled analogues of citalopram as potential radioligands for the serotonin transporter. Bioorganic & Medicinal Chemistry. 2003;**11**:3447-3456. DOI: 10.1016/j.bmc.2009.05.021

[44] Samuelsson L, Långström B. Synthesis of 1-(2′-deoxy-2′-fluoro-β-D-arabinofuranosyl)-[*methyl*-^{11}C]thymine ([^{11}C]FMAU) *via* a Stille cross-coupling reaction with [^{11}C]methyl

iodide. Journal of Labelled Compounds and Radiopharmaceuticals. 2003;**46**:263-272. DOI: 10.1002/jlcr.668

[45] Smith PJ, editor. Chemistry of TIN, UK: Blackie Academic & Professional; 1998. pp. 429-441. DOI: 10.1007/978-94-011-4938-9

[46] Prabhakaran J, Majo VJ, Simpson NR, Van Heertum RL, Mann JJ, Kumar JSD. Synthesis of [^{11}C]celecoxib: A potential PET probe for imaging COX-2 expression. Journal of Labelled Compounds and Radiopharmaceuticals. 2005;**48**:887-895. DOI: 10.1002/jlcr.1002

[47] Takashima-Hirano M, Takashima T, Katayama Y, Wada Y, Sugiyama Y, Watanabe Y, Doi H, Suzuki M. Efficient sequential synthesis of PET probes of the COX-2 inhibitor [^{11}C]celecoxib and its major metabolite [^{11}C]SC-62807 and *in vivo* PET behavior evaluation. Bioorganic & Medicinal Chemistry. 2011;**19**:2997-3004. DOI: 10.1016/j.bmc.2011.03.020

[48] Takashima T, Wu C, Takashima-Hirano M, Katayama Y, Wada Y, Suzuki M, Kusuhara H, Sugiyama Y, Watanabe Y. Evaluation of BCRP function in hepatobiliary and renal excretion using PET with [^{11}C]SC-62807. Journal of Nuclear Medicine. 2013;**54**:267-276. DOI: 10.1124/jpet.110.170092

[49] Suzuki M, Noyori R, Långström B, Watanabe Y. Molecular design of prostaglandin probes in brain research: High, specific binding to a novel prostacyclin receptor in the central nervous system. Bulletin of the Chemical Society of Japan. 2000;**73**:1053-1070. DOI: 10.1246/bcsj.73.1053

[50] Suzuki M, Doi H, Kato K, Björkman M, Långström B, Watanabe Y, Noyori R. Rapid methylation for the synthesis of a ^{11}C-labeled tolylisocarbacyclin imaging the IP_2 receptor in a living human brain. Tetrahedron. 2000;**56**:8263-8273. DOI: 10.1016/S0040-4020(00)00734-1

[51] Fowler JS, Wolf AP. Working against time: Rapid radiotracer synthesis and imaging the human brain. Accounts of Chemical Research. 1997;**30**:181-188. DOI: 10.1021/ar960068c

[52] Cui Y, Takamatsu H, Kakiuchi T, Ohba H, Kataoka Y, Yokoyama C, Onoe H, Watanabe Y, Hosoya T, Suzuki M, Noyori R, Tsukada H, Watanabe Y. Neuroprotection by a central nervous system-type prostacyclin receptor ligand demonstrated in monkeys subjected to middle cerebral artery occlusion and reperfusion: A positron emission tomography study. Stroke. 2006;**37**:2830-2836. DOI: 10.1161/01.STR.0000245088.60282.22

[53] Takashima T, Nagata H, Nakae T, Cui Y, Wada Y, Kitamura S, Doi H, Suzuki M, Maeda K, Kusuhara H, Sugiyama Y, Watanabe Y. Positron emission tomography studies using (15*R*)-16-*m*-[^{11}C]tolyl-17,18,19,20-tetranorisocarbacyclin methyl ester for the evaluation of hepatobiliary transport. Journal of Pharmacology and Experimental Therapeutics. 2010;**335**:314-323. DOI: 10.1124/jpet.110.170092; Takashima T, Kitamura S, Wada Y, Tanaka M, Shigihara Y, Ishii H, Ijuin R, Shiomi S, Nakae T, Watanabe Y, Cui Y, Doi H, Suzuki M, Maeda K, Kusuhara H, Sugiyama Y, Watanabe Y. Positron emission

tomography (PET) imaging-based evaluation of hepatobiliary transport in humans using 15*R*-[^{11}C]TIC-Me. The Journal of Nuclear Medicine. 2012;**53**:741-748. DOI: 10.2967/jnumed.111.098681

[54] Björkman M, Doi H, Resul B, Suzuki M, Noyori R, Watanabe Y, Långström B. Synthesis of a ^{11}C-labelled prostaglandin $F_{2\alpha}$ analogue using an improved method for Stille reactions with [^{11}C] methyl iodide. Journal of Labelled Compounds and Radiopharmaceuticals. 2000;**43**:1327-1334. DOI: 10.2967/jnumed.111.098681

[55] Suzuki M, Sumi K, Koyama H, Siqin, Hosoya T, Takashima-Hirano M, Doi H. Pd(0)-mediated rapid coupling between methyl iodide and hetero-arylstannanes: An efficient and general method for the incorporation of a positron-emitting ^{11}C radionuclide into heteroaromatic frameworks. Chemistry - A European Journal. 2009;**15**:12489-12495. DOI: 10.1002/chem.200901145

[56] Littke AF, Schwarz L, Fu GC. Pd/P(*t*-Bu)$_3$: A mild and general catalyst for Stille reactions of aryl chlorides and aryl bromides. Journal of the American Chemical Society. 2002;**124**:6343-6348. DOI: 10.1021/ja020012f

[57] Suzuki M, Takashima-Hirano M, Koyama H, Yamaoka T, Sumi K, Nagata H, Hidaka H, Doi H. Efficient synthesis of [^{11}C]H-1152, a PET probe specific for Rho-kinases, highly potential targets in diagnostic medicine and drug development. Tetrahedron. 2012;**68**:2336-2341. DOI: 10.1016/j.tet.2012.01.033

[58] Reich HJ, Green DP, Medina MA, Goldenberg WS, Gudmundsson BÖ, Dykstra RR, Phillips NH. Aggregation and reactivity of phenyllithium solutions. Journal of the American Chemical Society. 1998;**120**:7201-7210. DOI: 10.1021/ja980684z

[59] Bading JR, Shields AF. Imaging of cell proliferation: Status and prospects. Journal of Nuclear Medicine. 2008;**49**:64S-80S. DOI: 10.2967/jnumed.107.046391

[60] Lee TS, Ahn SH, Moon BS, Chun KS, Kang JH, Cheon GJ, Choi CW, Lim SM. Comparison of ^{18}F-FDG, ^{18}F-FET and ^{18}F-FLT for differentiation between tumor and inflammation in rats. Nuclear Medicine and Biology. 2009;**36**:681-686. DOI: 10.1016/j.nucmedbio.2009.03.009

[61] Toyohara J, Okada M, Toramatsu C, Suzuki K, Irie T. Feasibility studies of 4′-[methyl-^{11}C]thiothymidine as a tumor proliferation imaging agent in mice. Nuclear Medicine and Biology. 2008;**35**:67-74. DOI: 10.1016/j.nucmedbio.2007.10.001

[62] Koyama H, Siqin, Zhang Z, Sumi K, Hatta Y, Nagata H, Doi H, Suzuki M. Highly efficient syntheses of [methyl-^{11}C]thymidine and its analogue 4′-[methyl-^{11}C]thiothymidine as nucleoside PET probes for cancer cell proliferation by Pd^0-mediated rapid *C*-[^{11}C]methylation. Organic & Biomolecular Chemistry. 2011;**9**:4287-4294. DOI: 10.1039/c0ob01249a

[63] Zhang Z, Doi H, Koyama H, Watanabe Y, Suzuki M. Efficient syntheses of [^{11}C]zidovudine and its analogues by convenient one-pot palladium(0)-copper(I) co-mediated rapid *C*-[^{11}C]methylation. Journal of Labelled Compounds and Radiopharmaceuticals. 2014;**57**:540-549. DOI: 10.1002/jlcr.3213

[64] Tahara T, Zhang Z, Ohno M, Hirao Y, Hosaka N, Doi H, Suzuki M, Onoe H. A novel ^{11}C-labeled thymidine analog, [^{11}C]AZT, for tumor imaging by positron emission tomography. EJNMMI Research. 2015;**5**:45. DOI: 10.1186/s13550-015-0124-0

[65] Hosoya T, Sumi K, Doi H, Wakao M, Suzuki M. Rapid methylation on carbon frameworks useful for the synthesis of $^{11}CH_3$-incorporated PET tracers: Pd(0)-mediated rapid coupling of methyl iodide with an alkenyltributylstannane leading to a 1-methylalkene. Organic & Biomolecular Chemistry. 2006;**4**:410-415. DOI: 10.1039/b515215a

[66] Menzel K, Fu GC. Room-temperature Stille cross-couplings of alkenyltin reagents and functionalized alkyl bromides that possess β hydrogens. Journal of the American Chemical Society. 2003;**125**:3718-3719. DOI: 10.1021/ja0344563

[67] Mee SPH, Lee V, Baldwin JE. Stille coupling made easier—The synergic effect of copper (I) salts and the fluoride ion. Angewandte Chemie International Edition. 2004;**43**:1132-1136. DOI: 10.1002/anie.200352738

[68] Hosoya T, Wakao M, Kondo Y, Doi H, Suzuki M. Rapid methylation of terminal acetylenes by the Stille coupling of methyl iodide with alkynyltributylstannanes: A general protocol potentially useful for the synthesis of short-lived $^{11}CH_3$-labeled PET tracers with a 1-propynyl group. Organic & Biomolecular Chemistry. 2004;**2**:24-27. DOI: 10.1039/B311532A

[69] Skuballa W, Vorbrüggen H. A new route to 6α-carbacyclins-synthesis of a stable, biologically potent prostacyclin analogue. Angewandte Chemie International Edition. 1981;**20**:1046-1048. DOI: 10.1002/anie.198110461

[70] Miyaura N, Suzuki A. Palladium-catalyzed cross-coupling reactions of organoboron compound. Chemical Reviews. 1995;**95**:2457-2483. DOI: 10.1021/cr00039a007

[71] Hostetler ED, Terry GE, H. D. Burns HD. An improved synthesis of substituted [^{11}C] toluenes via Suzuki coupling with [^{11}C]methyl iodide. Journal of Labelled Compounds and Radiopharmaceuticals. 2005;**48**:629-634. DOI: 10.1002/jlcr.935

[72] Doi H, Ban I, Nonoyama A, Sumi K, Kuang C, Hosoya T, Tsukada H, Suzuki M. Palladium(0)-mediated rapid methylation and fluoromethylation on carbon frameworks by reacting methyl and fluoromethyl iodide with aryl and alkenyl boronic acid esters useful for the synthesis of C–$^{11}CH_3$ and C–$^{18}FCH_2$ containing PET tracers. Chemistry - A European Journal. 2009;**15**:4165-4171. DOI: 10.1002/chem.200801974

[73] Kanazawa M, Furuta K, Doi H, Mori T, Minami T, Ito S, Suzuki M. Synthesis of an acromelic acid A analog-based ^{11}C-labeled PET tracer for exploration of the site of action of acromelic acid A in allodynia induction. Bioorganic & Medicinal Chemistry Letters. 2011;**21**:2017-2020. DOI: 10.1016/j.bmcl.2011.02.018

[74] Suzuki M, Takashima-Hirano M, Ishii H, Watanabe C, Sumi K, Koyama H, Doi H. Synthesis of ^{11}C-labeled retinoic acid, [^{11}C]ATRA, via an alkenylboron precursor by Pd (0)-mediated rapid C-[^{11}C]methylation. Bioorganic & Medicinal Chemistry Letters. 2014;**24**:3622-3625. DOI: 10.1016/j.bmcl.2014. 05.041

[75] Shitara Y, Horie T, Eur SY. Transporters as a determinant of drug clearance and tissue distribution. Journal of Pharmaceutical Sciences. 2006;**27**:425-446. DOI: 10.1016/j.ejps.2005.12.003

[76] Shepherd J, Cobbe SM, Ford I, Isles CG, Lorimer AR, MacFarlane PW, McKillop JH, Packard CJ. Prevention of coronary heart disease with pravastatin in men with hypercholesterolemia. The New England Journal of Medicine. 1995;**333**:1301-1307. DOI: 10.1056/NEJM199511163332001

[77] Ijuin R, Takashima T, Watanabe Y, Sugiyama Y, Suzuki M. Synthesis of [^{11}C] dehydropravastatin, a PET probe potentially useful for studying OATP1B1 and MRP2 transporters in the liver. Bioorganic & Medicinal Chemistry. 2012;**20**:3703-3709. DOI: 10.1016/j.bmc.2012.04.051

[78] Shingaki T, Takashima T, Ijuin R, Zhang X, Onoue T, Katayama Y, Okauchi T, Hayashinaka E, Cui Y, Wada Y, Suzuki, Maeda K, Kusuhara H, Sugiyama Y, Watanabe Y. Evaluation of OATP and MRP2 activities in hepatobiliary excretion using newly developed positron emission tomography tracer [^{11}C]dehydropravastatin in rats. Journal of Pharmacology and Experimental Therapeutics. 2013;**347**:193-202. DOI: 10.1124/jpet.113.206425

[79] Frisch AC, Beller M. Catalysts for cross-coupling reactions with non-activated alkyl halides. Angewandte Chemie International Edition. 2005;**44**:674-688. DOI: 10.1002/anie.200461432

[80] Ishiyama T, Abe S, Miyaura N, Suzuki A. Palladium-catalyzed alkyl-alkyl cross-coupling reaction of 9-alkyl-9-bbn derivatives with iodoalkanes possessing β-hydrogens. Chemistry Letters. 1992:691-694. DOI: 10.1246/cl.1992.691

[81] Netherton MR, Dai C, Neuschütz K, Fu GC. Room-temperature alkyl–alkyl Suzuki cross-coupling of alkyl bromides that possess β hydrogens. Journal of the American Chemical Society. 2001;**123**:10099-10100. DOI: 10.1021/ja011306o

[82] Koyama H, Zhan Z, Ijuin R, Siqin, Son J, Hatta Y, Ohta M, Wakao M, Hosoya T, Doi H, Suzuki M. Pd^0-mediated rapid coupling of methyl iodide with excess amounts of benzyl- and cinnamylboronic acid esters: Efficient method for incorporation of positron-emitting ^{11}C radionuclide into organic frameworks by coupling between two sp^3-hybridized carbons. RSC Advances. 2013;**3**:9391-9401. DOI: 10.1039/c3ra40815a

[83] Allman T, Goel RG. The basicity of phosphines. Canadian Journal of Chemistry. 1982;**60**:716-722. DOI: 10.1139/v82-106

[84] Koyama H, Doi H, Suzuki M. Evaluation of TlOH effect for Pd^0-mediated cross-coupling of methyl iodide and excess boronic acid ester toward fabrication of [^{11}C] CH_3-incorporated PET tracer. International Journal of Organic Chemistry. 2013;**3**:220-223. DOI: 10.4236/ijoc.2013.33030

[85] Uenishi J, Beau J-M, Armstrong RW, Kishi Y. Dramatic rate enhancement of Suzuki diene synthesis. Its application to palytoxin synthesis. Journal of the American Chemical Society. 1987;**109**:4756-4758. DOI: 10.1021/ja00249a069

[86] Doi H, Goto M, Suzuki M. Pd^0-mediated rapid *C*-[^{18}F]fluoromethylation by the cross-coupling reaction of a [^{18}F]fluoromethyl halide with an arenylboronic acid ester: Novel method for the synthesis of a ^{18}F-labeled molecular probe for positron emission tomography. Bulletin of the Chemical Society of Japan. 2012;**85**:1233-1238. DOI: 10.1246/bcsj.20120151

[87] Itsenko O, Kihlberg T, Långström B. Labeling of aliphatic carboxylic acids using [^{11}C] carbon monoxide. Nature Protocols. 2006;**1**:798-802. DOI: 10.1038/nprot.2006.112

[88] Kealey S, Plisson C, Collier TL, Long Nicholas J, Husbands SM, Martarello L, Gee AD. Microfluidic reactions using [^{11}C]carbon monoxide solutions for the synthesis of a positron emission tomography radiotracer. Organic & Biomolecular Chemistry. 2011;**9**:3313-3319. DOI: 10.1039/c0ob00631a

[89] Eriksson J, Johan H, Windhorst AD. Transition metal mediated synthesis using [^{11}C]CO at low pressure—A simplified method for ^{11}C-carbonylation. Journal of Labelled Compounds and Radiopharmaceuticals. 2012;**55**:223-228. DOI: 10.1002/jlcr.2930

[90] Yamamoto Y. The first general and selective palladium(II)-catalyzed alkoxycarbonylation of arylboronates: Interplay among benzoquinone-ligated palladium(0) complex, organoboron, and alcohol solvent. Advanced Synthesis and Catalysis. 2010;**352**:478-492. DOI: 10.1002/adsc.200900836

[91] Takashima-Hirano M, Ishii H, Suzuki M. Synthesis of [^{11}C]Am80 via novel Pd(0)-mediated rapid [^{11}C]carbonylation using arylboronate and [^{11}C]carbon monoide. ACS Medicinal Chemical Letters. 2012;**3**:804-807. DOI: 10.1021/ml300160w

[92] Ishii H, Minegishi K, Nagatsu K, Zhang M-R. Pd(0)-mediated [^{11}C]carbonylation of aryl and heteroaryl boronic acid pinacol esters with [^{11}C]carbon monoxide under ambient conditions and a facile process for the conversion of [*carbonyl*-^{11}C]esters to [*carbonyl*-^{11}C] amides. Tetrahedron. 2015;**71**:1588-1596. DOI: 10.1016/j.tet.2015.01.008

[93] Dahl K, Schou M, Amini N, Halldin C. Palladium-mediated [^{11}C]carbonylation at atmospheric pressure: A general method using xantphos as supporting ligand. European Journal of Organic Chemistry. 2013;1228-1231. DOI: 10.1002/ejoc.201201708

[94] Rahman O, Långström B, Halldin C. Alkyl iodides and [^{11}C]CO in nickel-mediated cross-coupling reactions: Successful use of alkyl electrophiles containing a β hydrogen atom in metal-mediated [^{11}C]carbonylation. Chemistry Select. 2016;**1**:2498-2501. DOI: 10.1002/slct.201600643

[95] Kealey S, Gee A, Miller PW. (Review) Transition metal mediated [^{11}C]carbonylation reactions: Recent advances and applications. Journal of Labelled Compounds and Radiopharmaceuticals. 2014;**57**:195-201. DOI: 10.1002/jlcr.3150

[96] Riss PJ, Lu S, Telu S, Aigbirhio FI, Pike VW. Cu^I-catalyzed ^{11}C carboxylation of boronic acid esters: A rapid and convenient entry to ^{11}C-labeled carboxylic acids, esters, and amides. Angewandte Chemie International Edition in English. 2012;**51**:2698-2702. DOI: 10.1002/anie.201107263

[97] Goto M, Kurai S, Yamamoto R, Kanamitsu T, Suzuki M, Watanabe Y, Doi H. In: 2015 International Chemical Congress of Pacific Basin Societies; 15–20 December, 2015; Honolulu, Hawaii. 2015. #227

[98] Takashima-Hirano M, Shukuri M, Takashima T, Goto M, Wada Y, Watanabe Y, Onoe H, Doi H, Suzuki M. General method for ^{11}C-labeling of 2-arylpropionic acids and their esters: Construction of a PET tracer library for a study of biological events involved in COXs expression. Chemistry - A European Journal. 2010;**16**:4250-4258. DOI: 10.1002/chem.200903044

[99] Shukuri M, Mawatari A, Ohno M, Suzuki M, Doi H, Watanabe Y, Onoe H. Detection of cyclooxygenase-1 in activated microglia during amyloid plaque progression: PET studies in Alzheimer's disease model mice. Journal of Nuclear Medicine. 2016;**57**:291-296. DOI: 10.2967/jnumed.115.166116

[100] Ohnishi A, Senda M, Yamane T, Sasaki M, Mikami T, Nishio T, Ikari Y, Nishida H, Shukuri M, Takashima T, Mawatari A, Doi H, Watanabe Y, Onoe H. Human whole-body biodistribution and dosimetry of a new PET tracer, [^{11}C]ketoprofen methyl ester, for imagings of neuroinflammation. Nuclear Medicine and Biology. 2014;**41**:594-599. DOI: 10.1016/j/nucmedbio.2014.04.008; Ohnishi A, Senda M, Yamane T, Mikami T, Nishida H, Nishio T, Akamatsu G, Ikari Y, Kimoto S, Aita K, Sasaki M, Shinkawa H, Yamamoto Y, Shukuri M, Mawatari A, Doi H, Watanabe Y, Onoe H. Exploratory human PET study of the effectiveness of ^{11}C-ketoprofen methyl ester, a potential biomarker of neuroinflammatory processes in Alzheimer's disease. Nuclear Medicine and Biology. 2016;**43**:438-444. DOI: 10.1016/j.nucmedbio.2016.04.005; Nozaki S, Ozaki N, Suzuki S, Goto M, Mawatari A, Nakatani Y, Hayashinaka E, Wada Y, Doi H, Watanabe Y. Development of diagnostic techniques for early rheumatoid arthritis using positron emission tomography with [^{11}C]PK11195 and [^{11}C]ketoprofen tracers. Molecular Imaging and Biology. 2017;**19**:746-753. DOI: 10.1007/s11307-016-1039-5

[101] Kato K, Kikuchi T, Nengaki N, Arai T, Zhang M-R. Tetrabutylammonium fluoride-promoted α-[^{11}C]methylation of α-arylesters: A simple and robust method for the preparation of ^{11}C-labeled ibuprofen. Tetrahedron Letters. 2010;**51**:5908-5911. DOI: 10.1016/j.tetlet.2010.09.007

[102] Link JM, Krohn KA, Clark JC. Production of [^{11}C]CH_3I by single pass reaction of [^{11}C]CH_4 with I_2. Nuclear Medicine and Biology. 1997;**24**:93-97. DOI: 10.1016/S0969-8051(96)00181-3

[103] Noguchi J, Suzuki K. Automated synthesis of the ultra high specific activity of [^{11}C]Ro15-4513 and its application in an extremely low concentration region to an ARG study. Nuclear Medicine and Biology. 2003;**30**:335-343. DOI: 10.1016/ S0969-8051(02)00422-5

[104] Doi H, Sato K, Shindou H, Sumi K, Koyama H, Hosoya T, Watanabe Y, Ishii S, Tsukada H, Nakanishi K, Suzuki M. Blood-brain barrier permeability of ginkgolide: Comparison of the behavior of PET probes 7α-[^{18}F]fluoro- and 10-*O*-*p*-[^{11}C]methylbenzyl ginkgolide B in monkey and rat brain. Bioorganic & Medicinal Chemistry. 2016;**24**:5148-5157. DOI: 10.1016/ j.bmc.2016.08.032

[105] Koyama H, Ikenuma H, Toda H, Kondo G, Hirano M, Kato M, Abe J, Yamada T, Wakabayashi T, Ito K, Natsume A, Suzuki M. Synthesis of PET probe O^6-[(3-[^{11}C]methyl) benzyl]guanine by Pd^0-mediated rapid *C*-[^{11}C]methylation toward imaging DNA repair protein O^6-methylguanine-DNA methyltransferase in glioblastoma. Bioorganic & Medicinal Chemistry Letters. 2017;**27**:1892-1896. DOI: 10.1016/j.bmcl.2017.03.045

[106] Goto M, Mizuma H, Wada Y, Onoe H, Watanabe Y, Suzuki M, Doi H. In: 52nd Symposium on the Chemistry of Natural Products. Abstract in English, see Symposium Paper; September 29–October 1, 2010; Shizuoka, Japan. 2010. p. 264

[107] Takahashi K, Hosoya T, Onoe K, Doi H, Nagata H, Hiramatsu T, Li X-L, Watanabe Yu, Wada Y, Takashima T, Suzuki M, Onoe H, Watanabe Y. [^{11}C]Cetrozole: An improved C-[^{11}C]methylated PET probe for aromatase imaging in the brain. Journal of Nuclear Medicine. 2014;**55**:852-857. DOI: 10.2967/jnumed.113.131474. Patent: JP2009-159129

[108] Doi H, Mawatari A, Kanazawa M, Nozaki S, Nomura Y, Kitayoshi T, Akimoto K, Suzuki M, Ninomiya S, Watanabe Y. Synthesis of ^{11}C-labeled thiamine and fursultiamine for *in vivo* molecular imaging of vitamin B_1 and its prodrug using positron emission tomography. The Journal of Organic Chemistry. 2015;**80**:6520-6258. DOI: 10.1021/acs.joc.5b00685. Patent: JP2012-47448

[109] Kimura T, Sako T, Siqi, Hosokawa-Muto J, Cui YL, Wada Y, Kataoka Y, Doi H, Sakaguchi S, Suzuki M, Watanabe Y, Kuwata K. Synthesis of an ^{11}C-labeled antiprion GN8 derivative and evaluation of its brain uptake by positron emission tomography. ChemMedChem. 2013;**8**:1035-1039. DOI: 10.1002/cmdc.201300167

[110] Tarkiainen J, Vercouillie J, Emond P, Sandell J, Hiltunen J, Frangin Y, Guilloteau D, Halldin C. Carbon-11 labelling of MADAM in two different positions: A highly selective PET radioligand for the serotonin transporter. Journal of Labelled Compounds and Radiopharmaceuticals. 2001;**44**:1013-1023. DOI: 10.1002/jlcr.523

[111] Gerasimov MR, Ferrieri RA, Schiffer WK, Logan J, Gatley SJ, Gifford AN, Alexoff DA, Marsteller DA, Shea C, Garza V, Carter P, King P, Ashby CR Jr, Vitkun S, Dewey SL. Study of brain uptake and biodistribution of [^{11}C]toluene in non-human primates and mice. Life Sciences. 2002;**70**:2811-2828. DOI: 10.1016/S0024-3205(02)01542-4

[112] Iida Y, Ogawa M, Ueda M, Tominaga A, Kawashima H, Magata Y, Nishiyama S, Tsukada H, Mukai T, Saji H. Evaluation of 5-^{11}C-methyl-A-85380 as an imaging agent for PET investigations of brain nicotinic acetylcholine receptors. Journal of Nuclear Medicine. 2004;**45**:878-884

[113] Langer O, Forngren T, Sandell J, Dollé F, Långström B, Någren K, Halldin C. Preparation of 4-[^{11}C]methylmetaraminol, a potential PET tracer for assessment of myocardial sympathetic innervation. Journal of Labelled Compounds and Radiopharmaceuticals. 2003;**46**:55-65. DOI: 10.1002/jlcr.642

[114] Ishiwata K, Kawamura K, Yajima K, QingGeLeTu, Mori H, Shiba K. Evaluation of (+)-*p*-[^{11}C]methylvesamicol for mapping sigma1 receptors: A comparison with [^{11}C]SA4503. Nuclear Medicine and Biology. 2006;**33**:543-548. DOI: 10.1016/j.nucmedbio.2006.01.008

[115] Kawamura K, Shiba K, Tsukada H, Nishiyama S, Mori H, Ishiwata K. Synthesis and evaluation of vesamicol analog (-)-*o*-[^{11}C]methylvesamicol as a PET ligand for vesicular acetylcholine transporter. Annals of Nuclear Medicine. 2006;**20**:417-424. DOI: 10.1007/BF03027377

[116] Bennacef I, Perrio C, Lasne M-C, Barré L. Functionalization through lithiation of (*S*)-*N*-(1-Phenylpropyl)-2-phenylquinoline-4-carboxamide. Application to the labeling with carbon-11 of NK-3 receptor antagonist SB 222200. Journal of Organic Chemistry. 2007;**72**:2161-2165. DOI: 10.1021/jo062285p

[117] Zeng F, Mun J, Jarkas N, Stehouwer JS, Voll RJ, Tamagnan GD, Howell L, Votaw JR, Kilts CD, Nemeroff CB, Goodman MM. Synthesis, radiosynthesis, and biological evaluation of carbon-11 and fluorine-18 labeled reboxetine analogues: Potential positron emission tomography radioligands for in vivo imaging of the norepinephrine transporter. Journal of Medicinal Chemistry. 2009;**52**:62-73. DOI: 10.1021/jm800817h

[118] Toyohara J, Sakata M, Wu J, Ishikawa M, Oda K, Ishii K, Iyo M, Hashimoto K, Ishiwata K. Preclinical and the first clinical studies on [^{11}C]CHIBA-1001 for mapping α7 nicotinic receptors by positron emission tomography. Annals of Nuclear Medicine. 2009;**23**:301-309. DOI: 10.1007/s12149-009-0240-x

[119] Kawamura K, Yui J, Konno F, Yamasaki T, Hatori A, Wakizaka H, Fujinaga M, Kumata K, Yoshida Y, Ogawa M, Nengaki N, Yanamoto K, Fukumura T, Zhang M-R. Synthesis and evaluation of PET probes for the imaging of I2 imidazoline receptors in peripheral tissues. Nuclear Medicine and Biology. 2012;**39**:89-99. DOI: 10.1016/j.nucmedbio.2011.06.001

[120] Fujinaga M, Yamasaki T, Maeda J, Yui J, Xie L, Nagai Y, Nengaki N, Hatori A, Kumata K, Kawamura K, Zhang M-R. Development of N-[4-[6-(isopropylamino)pyrimidin-4-yl]-1,3-thiazol-2-yl]-N-methyl-4-[^{11}C]methylbenzamide for positron emission tomography imaging of metabotropic glutamate 1 receptor in monkey brain. Journal of Medicinal Chemistry. 2012;**55**:11042-11051. DOI: 10.1021/jm301597s

[121] Andersen VL, Herth MM, Lehel S, Knudsen GM, Kristensen JL. Palladium-mediated conversion of para-aminoarylboronic esters into para-aminoaryl-^{11}C-methanes. Tetrahedron Letters. 2013;**54**:213-216. DOI: 10.1016/j.tetlet.2012.11.001

[122] Nishiyama S, Ohba H, Kobashi T, Nakamasu Y, Nakao H, Ogata T, Kitashoji T, Tsukada H. Development of novel PET probe [^{11}C](R,R)HAPT and its stereoisomer [^{11}C](S,S) HAPT for vesicular acetylcholine transporter imaging: A PET study in conscious monkey. Synapse. 2014;**68**:283-292. DOI: 10.1002/syn.21743

[123] Kanazawa M, Ohba H, Iwazaki A, Kakiuchi T, Tsukada H. Synthesis of 6-[^{11}C]methyl-*m*-tyrosine ([^{11}C]6MemTyr) for dopamine synthesis imaging in living brain using PET. Bioorganic & Medicinal Chemistry. 2015;**23**:729-734. DOI: 10.1016/j.bmc.2014.12.061

[124] Fujinaga M, Yamasaki T, Nengaki N, Ogawa M, Kumata K, Shimoda Y, Yui J, Xie L, Zhang Y, Kawamura K, Zhang M-R. Radiosynthesis and evaluation of 5-methyl-N-(4-[^{11}C]methylpyrimidin-2-yl)-4-(1H-pyrazol-4-yl)thiazol-2-amine ([^{11}C]ADX88178) as a

novel radioligand for imaging of metabotropic glutamate receptor subtype 4 (mGluR4). Bioorganic & Medicinal Chemistry Letters. 2016;**26**:370-374. DOI: 10.1016/j.bmcl.2015.12.008

[125] Rowland M. For recent commentaries for microdosing see: [Europe] The European Medicines Agency, Evaluation of Medicines for Human Use (EMEA), Position paper on non-clinical safety studies to support clinical trials with a single microdose, CPMP/SWP/2599/02/Rev1, 23 June 2004; [USA] U.S. Department of Health and Human Services, Food and Drug Administration, Center for Drug Evaluation and Research, Guidance for Industry, Investigators, and Reviewers, Exploratory IND Studies, January 2006; [Japan] Japan's response to microdosing study, Notification No. 0603001, 3 June 2008, Guidance for the Conduct of Microdosing Clinical Trials, Evaluation and Licensing Division, Pharmaceutical and Food Safety Bureau, Ministry of Health, Labour and Welfare, Japan. Journal of Pharmaceutical Sciences. 2012;**101**:4067-4074